MAINE TO GREENLAND

EXPLORING THE MARITIME FAR NORTHEAST

MAINE

TO GREENLAND

EXPLORING THE MARITIME FAR NORTHEAST

William W. Fitzhugh
Wilfred E. Richard

Photographs by Wilfred E. Richard

Smithsonian Books

WASHINGTON, DC

CONTENTS

Maine to Greenland: Exploring the Maritime Far Northeast
was produced by Perpetua Press, Santa Barbara, CA
Edited by Letitia O'Connor
Designed by Dana Levy

Published by Smithsonian Books
Director: Carolyn Gleason
Production Editor: Christina Wiginton

Library of Congress Cataloging-in-Publication Data

Fitzhugh, William W., 1943- author; Richard, Wilfred E., 1940-author

Maine to Greenland : exploring the maritime far northeast / William W. Fitzhugh and Wilfred E. Richard ; photographs by Wilfred E. Richard.

    pages cm

  Includes bibliographical references and index.

  ISBN 978-1-58834-377-2

  1. Atlantic Provinces--Description and travel. 2. Atlantic Provinces--Pictorial works. 3. Maine--Description and travel. 4. Maine--Pictorial works. 5. Greenland--Description and travel. 6. Greenland--Pictorial works. 7. Human ecology--Atlantic Provinces. 8. Human ecology--Maine. 9. Human ecology--Greenland. I. Richard, Wilfred, author, photographer. II. Title.

  F1035.8.F47 2013

  917.15--dc23

                    2013010782

Printed in China through Oceanic Graphic Printing, not at government expense

18 17 16 15 14   5 4 3 2 1

For permission to reproduce illustrations appearing in this book, please correspond directly with the owners of the works, identified at right. Smithsonian Books does not retain reproduction rights for these images individually, or maintain a file of addresses for sources.

P. 1: Arctic Poppies
The Moravians brought poppies (Papaver family) from Europe to their mission stations along the Labrador coast in late 1700s and early 1800s to brighten their lives and remind them of home. Here they thrive around the Moravian churchyard in Nain.

P. 2-3: Uummannaq annual dog sled race

Text © 2014 by Smithsonian Institution and Wilfred E. Richard

Images © 2014 by Wilfred E. Richard except as listed below.

All photographic images in this book were taken by, and rights are held by, Wilfred E. Richard, except as indicated below. The large format maps were created by Richard D. Kelly, Jr., and Marcia Bakry (Smithsonian Department of Anthropology) except for p. 56, 3.02 which is solely Bakry.

William W. Fitzhugh: p. 10, p. 20 : 1.11, p. 84 : 4.05, 4.06, 4.07, p. 99 : 4.28, p. 101: 4.32, p. 113 : 5.21, p. 113 : 5.23, p. 120 : 5.37, 5.38, p. 126 : 6.04, p. 127 : 6.05, 6.06, 6.07, p. 129 : 6.10 , p. 132 : 6.14, 6.15, p. 133 : 6.16, 6.17, 6.18, p. 134 : 6.19, p. 135 : 6.20, 6.21, p. 136 : 6.22, 6.23, 6.24, p. 137 : 6.25, p. 143 : 6.34, p. 145 : 6.36, 6.37, p. 146 : 6.38, p. 148: 6.41, p. 151 : 6.44, p. 158 : 7.06, p. 166 : 7.21, p. 167 : 7.23, 7.24, p. 241 : 10.03, p. 242 : 10.04

Stephen Loring: p. 18: 1.08, p. 44 : 2.15, p. 125 : 6.03, p. 128 : 6.08, p. 130 : 6.11, p. 138 : 6.26, p. 141 : 6.31, p. 147, 6.39.

Peary Archives, Peary-MacMillan Museum, Bowdoin College: p. 30: 1.22, 1.23, 1.24, p 31: 1.25, 1.26, 1.27.

Arctic Council and the International Science Committee (IASC) - *Arctic Climate Impact Assessment Report, 2004*: p. 36: 2.02, 2.03, p. 38: 2.06, p. 39: 2.08

National Snow and Sea Ice Data Center, 2004: p. 39: 2.07

Fisheries and Oceans Canada: p. 44: 2.15

Walter Adey and L-A. Hayek: p. 53: 2.32

Nick Caloyianis: p. 52: 2.30

Karen Loveland: p. 52: 2.31, p. 53: 2.33

Eric Phaneuf: p. 72: 3.32, p. 119: 5.32

Maine State Museum: p. 73: 3.35

Stearns A. Morse: p. 86: 4.09, 4.10Bill Duffy: p. 92: 4.20

Parks Canada: p. 99: 4.29, p. 100: 4.30Henry Youle Hind: p. 130: 6.12 [this is from his book, cited in bibiog]

National Museum of Natural History, Smithsonian Institution, Don Hurlbut: p. 131: 6.13

Institute of Cultural and Social Anthropology, Georg-August University: p. 138: 6.27

Angsar Walk: p. 159: 7.07, 7.08

John White illustration, The British Museum: p. 165: 7.20

Charles Francis Hall, engraving from his 1865 book: p. 166: 7.22.

Paul Nicklen, National Geographic Society: p. 205: 9.04

Inger Knudsen Holm Collection: p. 209: 9.10, 9.11Rockwell Kent watercolor and pen-and-ink drawing, p. 222: 9.32, 9.33: Bowdoin College Museum of Art, Brunswick, Maine Museum Purchase with Funds Donated Anonymously & Plattsburgh State Art Museum, Rockwell Kent Gallery and Collection. Digital photography by Peter Siegel.

Abraham Anghik Ruben sculpture, photo by Kipling Gallery, Toronto: p. 243: 10.05

National Research Council: p. 239: 10.01

Two systems of measurement are used in the region designated as the Maritime Far Northeast. In Canada and Greenland the metric system is used, while in the United States the English system is used. In this work, most measurements are presented in both metric and English systems. Conversion values are as follows:

DISTANCE: Mile to kilometer multiply by 1.61; Kilometer to mile multiply by 0.62. One statute or land mile equals 1.15 nautical or geographical miles; nautical/geographic miles are converted to statute miles.

LENGTH: Foot to meter, multiply by .30; meter to foot, multiply by 3.28

AREA: Square kilometer to square mile, divide by 2.59; square mile to square kilometer, multiply by 2.59.
1 Mile^2 = 2.59 km^2; 1 km^2 = 0.3861mi^2

TEMPERATURE: *Centigrade to Fahrenheit multiply degree centigrade by 1.8*

WEIGHT: Kilograms to pounds, multiply by 2.21

Down how many roads among the stars must man propel himself in search of the final secret? …we have joined the caravan, you might say, at a certain point; we will travel as far as we can, but we cannot in one lifetime see all that we would like to see or learn all that we hunger to know… I can only say that here is a bit of my personal universe, the universe traversed in a long and uncompleted journey.

Loren Eiseley, *The Immense Journey*

*To my father, Emery G. Richard, and mother, Marie Sorrell Richard,
uncle, Wilfred C. Richard, aunt, Ethel Sorrell Reals
and the mentors of my teenage years,
Georges Bourque, Normand Paquette, Gene Tucker*

*To my wife, Lynne Fitzhugh; to my mentors, Elmer Harp, Jr., Tony Morse,
and Tony Williamson; to my many students and colleagues;
and to my friends throughout
the Maritime Far Northeast, all of whom made
my research possible, exciting, and productive*

PREFACE

THE COLLECTION OF JOURNEYS presented in this book in images and words celebrates a region that the authors have explored throughout their entire lives. Bill Fitzhugh, a Smithsonian anthropologist, found a fascination with the North at Dartmouth College, where he met the famous Arctic explorer and author Vilhjalmur Stefansson, and trained with Elmer Harp, who took him on his first archaeological dig at a Dorset site in Newfoundland. As a Harvard graduate student, he again joined an expedition led by Harp to the east coast of Hudson Bay. He has conducted four decades of archaeological research from Newfoundland to Baffin Island, starting in Labrador in the early 1970s. The mystery of the first European settlement in the New World, which was identified at L'Anse aux Meadows on the Newfoundland coast in 1960, resonated with his Danish Viking ancestry, which resulted in *Vikings: The North Atlantic Saga* (Fitzhugh and Ward 2000).

Geographer and photographer Will Richard grew up in New Hampshire and settled in Maine, New England states that served as his springboard to the North. In the mid-1990s he became a Registered Maine Guide and established an outfitting company featuring backcountry northern expeditions. As Will began tracing his French Canadian roots back to Quebec and Acadia, his journeys expanded northward and took on a life of their own. After a trip to Labrador and subsequently meeting Bill, he joined the Smithsonian's Gateways Project, and the Maritime Far Northeast (MFNE)—its lands, flora, and animals—became a consuming focus of his photography and studies, leading him northward to Ellesmere in northern Nunavut and eventually to Greenland.

Each author developed a fascination for northern life and its straightforward physical challenges: the seasons of long light and deep darkness and the cold of altitude and latitude. The North became a touchstone that has given their lives meaning by awakening their senses to explore its landscapes, terrestrial and marine, and to understand its cultural heritage, seeking awareness and knowledge. The authors also share the conviction that the peoples and cultures of this region offer lessons that need to be considered by others as the challenges of climate and environmental change, not only in the North but also in the wider world, shift the parameters of understanding.

As readers follow their travels north—to the Canadian Maritimes, Newfoundland and northern Quebec, then to Labrador, Baffin, and Ellesmere Islands and Greenland—we encounter ways of living that are more grounded and sustainable than what is becoming the norm in more highly populated, consumer-driven southern parts of the world. We can all learn lessons from these northern peoples who have not yet destroyed their forests, polluted their waters, and run their wildlife into extinction or confined them to refuges and game parks. This book is about the Native peoples and local homegrown Euro-Americans of the North, their ways of life, and the lands they have lived in for hundreds if not thousands of years. The story is heavily illustrated, for words alone can hardly describe the incredible landscapes we have seen, the human history uncovered, and the luminous personalities met along the way.

Will's photographs reveal the splendor of a complex region of plants, animals, lands, and cultures. Much in this northern landscape delights the eye and the spirit. Basalt or granite, gray to almost black, dominate the landscape. Adjacent to the mountainous coastal spine are littorals marked by rivers, peninsulas, and islands. Where there are rocks or trees, there are lichens, brilliantly colored communities of plants that transform the land. Nature's colonizers, lichens literally devour rock, producing the meager soil that turns up in patches here and there.

The sea, blue in the south but increasingly gray and wind-whipped to the north, often ice-covered in winter, connects life in the Maritime Far Northeast and is the economic basis for much of its settlement. High latitudes, along with weather and climate, collectively produce extended periods of darkness, bitter cold, and blasts of wind and snow. Yet during the dark hours of winter, people find time for family and friends, for reflection and storytelling. But, even here, people migrate to population centers—St. John's, Halifax, Nain, Iqaluit, and Nuuk—and traditional ways begin to fade and social stresses intensify.

In these lands many lessons come from the past.

The extinction of the Greenland Norse, who exceeded the carrying capacity of the land in good times and paid a price when the Little Ice Age arrived, and the near-extinction of medieval Icelanders, who overgrazed and lost their meager soil to erosion and extirpated their walrus, seals, and much of their fish, are cautionary tales (Fitzhugh and Ward 2000). On the other hand,

Outdoorsman
Will Richard (left) hiking in the northern Presidential Range of New Hampshire in early 1970s with Karl Keim, Hans and Manya Malmstedt, and Brendan, the wolfhound.

climate changes that are bad for some are good for others. The same medieval warming that helped the Norse cross the Atlantic and settle in Greenland made it possible for Alaska's Thule Eskimos (an early Inuit people) to expand into the Canadian Arctic and Greenland for whale hunting around AD 1200. When the onset of the Little Ice Age restricted whaling in northern waters it opened new grounds for hunting marine mammals in southern Greenland, Labrador, and Hudson Bay.

These journeys were not just about personal discoveries. We were looking for answers that could be applied to our rapacious consumption of the earth's most precious resources: water, hydrocarbons, grass for forage, forests, minerals, and soil. The small-scale societies that have lived and currently live in the North could be sources of inspiration for new ways of adapting to the south, not just because northern peoples have been geographically isolated and less affected by the world outside but because they have adapted to the resources at hand in a more sustainable manner, by using appropriate small-scale technology, heating with local materials like seal oil, the sun, or wood. Being attuned to the land, listening to its heartbeat, and with centuries of experience available in the wisdom of elders, northerners have created a different way of life.

Meeting these people, understanding their cultures, and sensing how they live with the land became the purpose of the ever-northward expanding journeys that are chronicled in this book.

Maine to Greenland begins with a geographical description of the Maritime Far Northeast, followed by a discussion of regionwide climate change and chapters, which progress northward from Maine to northern Greenland, that illustrate themes characteristic of the lands and peoples of the Maritime Far Northeast and their place in the wider world.

We have become familiar, over forty years of northern travels, with the normal year-to-year variation of climatic cycles and their effects. When Bill voyaged annually to Labrador or Baffin Island for archaeological surveys, some years the heavy pack ice would block his progress, sometimes for weeks, as easterly winds pushed the ice against the shore, preventing it from melting until mid-July. But the next summer the ice might be driven offshore by westerly winds in early June. Some years there would be terrible forest fires; others, nothing but rain or even periodic snow.

In recent years the pattern has changed, substantially and definitively enough that scientists identify it as "regime change." Noting a pronounced shift in climate, sea ice, and animal behavior in the 1990s, Canadian Inuit described these phenomenon by saying, "The world is faster now" (Krupnik and Jolly 2010). Corresponding changes have occurred in the Maritime Far Northeast over the same period: shorter winters, longer ice-free summers, unseasonable winter rain or summer snow in Labrador, and Christmases without shore-fast ice in northern Greenland. The disappearance of winter sea ice around Newfoundland and the Gulf of St. Lawrence resulted in catastrophic losses among ice-dependent harp seal pups, forecasting a major future decline of a species that has been a dependable harvest, sustaining Native and settler communities from Greenland to Newfoundland for hundreds if not thousands of years. We have tallied scores of such observations during our voyages to the Quebec Lower North shore in the last decade: dead baby harp seals on the rocks; failed berry harvests; and the appearance of new migrants, such as birds, sharks, sea turtles, and many others. Will encountered similar swings in climate and weather in his repeated visits to Baffin Island and Greenland, notably insufficient winter ice for safe travel by snowmobile. These firsthand observations are chronicled in the pages that follow, along with many human migrations and cultural changes caused by shifting climate and its impacts on the land, sea, and their biota that we have documented archaeologically.

For the world at large, but especially for the North—where a change in temperature of one single degree results in the remarkable transformation of ice to water or water to ice—the "big bear lurking in the room" is the dramatic decline of Arctic sea ice. Previously an esoteric subject, almost overnight climate has become a global issue of overwhelming importance to humanity. Arctic cultural history reveals how past

Dartmouth College Expedition Team
As a Harvard grad student, Bill Fitzhugh (far right) joined the archaeological team of his mentor, Dartmouth professor Elmer Harp Jr. (far left) on the east coast of Hudson Bay, in July and August 1967, gaining experience needed for his Labrador research. Other team members included (l-r) Douglas Harp, William McCarty, Dr. Jack Rinker, John Miksic, and William Caveney.

climate change has dramatically affected northern cultures, stimulating migrations, forcing new adaptations, and resulting in major shifts in animal and human populations. Understanding the history of the Maritime Far Northeast, whose peoples have adapted to climate change in the past and have much at stake in how it reshapes their homelands today, helps us envision solutions to climate and environmental challenges that will soon engage the entire world.

The history and economy of Maine, the southern anchor of the Maritime Far Northeast, illustrates biological, geographic, and cultural continuities that extend as far north as Greenland. Whaling and fishing, long an economic engine for Maine that connects it with the Maritime Far Northeast, is evident in its lighthouses and still-active traditions of boat building. Both of us began our journeys north from this place. As a student Bill built his own kayak, a type of boat invented and perfected by the Inuit for hunting and transport on rocky shores where there are few beaches for launching and landing. He was drawn to the 4,000-year prehistoric trade route of Labrador's Ramah chert—used for making stone tools and found as far south as southern New England—and visited the excavations at Naskeag Point where archaeologists uncovered evidence of Labrador and Newfoundland connections with Dorset Eskimo cultures and Viking voyages intermingled with local Native artifacts. Inspired by explorers Peary and MacMillan, Will began exploring the Gulf of Maine

and points north, heading toward the lands of his own forebears in the Canadian Maritimes and building the cross-border International Appalachian Trail. He still chooses to make his home in Maine.

The hearty individualists who forged the extension of the International Appalachian Trail (IAT) provide a new perspective on the Canadian Maritime Provinces, still best known from its troubled history of conflict between the French and English that was memorialized by Henry Wadsworth Longfellow's poem of 1847, "Evangeline, A Tale of Acadie." We trek through the landscape of Acadia, which is today called Nova Scotia, that Longfellow evoked in the opening of his tale:

> This is the forest primeval. The murmuring pines and the hemlocks,
> Bearded with moss, and in garments green, indistinct in the twilight…
> Loud from its rocky caverns, the deep-voiced neighboring ocean
> Speaks, and in accents disconsolate answers the wail of the forest.

An earlier French history is indicated by the place-name, Port Aux Basques, which in the sixteenth and seventeenth centuries teemed with Basque whalers and fishermen. Newfoundland, the southern base for nineteenth-century cod-fishing and sealing fleets, is also where Bill's experience in archaeology began. In excavating the 1,500-year-old Paleo-Eskimo Dorset site at Port au Choix, a traditional village where cod-fishing and seal-hunting persists, he discovered a passion for uncovering ancient history that gave focus to his life (Harp 2003).

History is the driving force onboard the Smithsonian research vessel *Pitsiulak*, which for the past three decades has voyaged north from Newfoundland to Labrador and Baffin Island to explore the Far Northeast's 10,000 years of ancient and modern Indian and Eskimo cultures. Over the last ten years the northern Gulf of St. Lawrence, the gateway region of North America, has been the focus of research by the Smithsonian's Arctic Studies Center, designed and directed by Bill; Will has chronicled this process as expedition photographer and fellow scribe. Home not only to some of the earliest European settlers but also a long-standing boundary between Inuit peoples of the Arctic and Indian tribes of the Subarctic, the Gulf of St. Lawrence marks the convergence of European and Native worlds, which led to complex historical and cultural relationships marked by both conflict and collaboration.

Labrador's long human history has made it a testing ground for the Smithsonian's scientific studies documenting the impact of climate change on the region's early cultures. Labrador became a vital center of an Inuit whaling culture from the sixteenth to the eighteenth centuries. The arrival of European traders and Moravian missionaries in the mid-1700s brought a clash of values between traditional Inuit religion and Christianity that eventually cost the Labrador Inuit their language, their traditional shamanistic religion, and key parts of their social and intellectual life. Similar damage has been inflicted on its Innu (Indian) peoples by the expansion of mining, dams, and Cold War fighter–pilot training. This conflict of values illustrates the power of outside forces on traditional societies in Labrador, Nunavut, and Greenland.

Greenland was the first of the MFNE lands to experience the collision of worlds: the western (European) and eastern (Inuit) branches of humanity first met here around AD 985 after migrating full circle around the globe. Norse settlers created a farming colony in Greenland that lasted 450 years before going extinct just a few decades before Columbus reached the Caribbean. Other than fleeting encounters ca. AD 1000 in Vinland, Markland, and the High Canadian Arctic, the first sustained contacts with New World peoples took place in the walrus- and whale-hunting grounds of Western Greenland's Disko Bay during the seventeenth century.

The joining of east and west begun by the Norse intensified after Nordic immigrants reappeared in Greenland in 1721. An extended period of paternalistic colonization produced a dual society in which many Greenlanders received European educations and broad benefits of Danish citizenship while others in more remote regions continued a more subsistence-based Inuit way of life, but one bolstered by a European-style safety net that is today also supported by modern technology. With nationhood looming, Greenland is poised to become autonomous, free for the first time to determine its policies and choose its own international partners.

Nunavut, the Inuit lands of the Eastern Arctic, had a very different history and cultural background. The Canadian Arctic was not settled by Europeans until after World War II, although contact with Yankee and Scottish whalers had major impacts from the 1830s to 1870s and at trading posts since the late nineteenth century. Political self-determination arrived in Nunavut in 1999 with its new status as an independent Canadian territory. The Inuit of Nunavut want to retain their language and traditions and continue to use subsistence resources. But the need for modernization and industrial development raise tough political and economic challenges to a region that did not emerge from a hunting-and-gathering way of life until the 1950s.

The political and economic situations of Greenland and Nunavut are similar but different; liberated from quasi-colonial status, both Inuit-dominated populations face choices about how to relate to their former host countries, to each other, and to the wider world. Warming climates and longer open-water seasons will bring new opportunities for trade and developing interlinked cultural, economic, and political paths to the future.

The cultural adaptations that have been developed to live in the cold, harsh region of the Far Northeast have also built strong community bonds. Although the coastal communities of Downeast Maine, the outports of Newfoundland and Labrador, and the settlements of Nunavut and Greenland all harbor poverty and material want, the traveler also quickly discovers vibrant life, a pragmatic optimism, friends in strangers, a lively dialogue, a kind word, a meal, and often a place to stay—sometimes for weeks. As a board member of the Maine Chapter of the International Appalachian Trail remarked:

> It is not necessarily those lands which are the most fertile or most favored in climate that seem to be the happiest, but those in which a long struggle of adaptation between man and his environment has brought out the best qualities in both (Urquhart 2004: 22).

Our intersection with these worlds comes from myriad directions, starting with our common Nordic backgrounds that were transplanted to the Americas. Our career trajectories have led us back to the northern lands our ancestors settled. Our searches— Bill's scientific quest to uncover the ancient history and cultures of this place and Will's explorations using photography and field observations—have culminated in this book, which we hope will bring the attention of a wider global audience to these regions at a time of great change.

Place and culture do shape who we are, our values, and ways of seeing. But after traveling and working in the North, we can no longer accept the conclusion that peoples of the temperate zone will necessarily derive the best solutions to society's problems. The diverse peoples, ancient or modern, who lived in the MFNE had one thing in common: to a major degree they remained attuned to the phenomena of nature. Unlike much of today's increasingly urbanized global culture, they are still intensely aware of their dependence on the natural forces of the earth—seasonal cycles, climate variability, changes in animal-migration patterns, and population fluctuations. The contrast between life practices among peoples in these northern lands and the more developed southern world is cause for reflection.

1.01 Summer Solstice on Baffin Bay

In the full light of summer solstice night, the landscape is reflected with Zenlike symmetry. The light that falls on and is reflected back by the unblemished whiteness of the Arctic is magnificent. It projects a spiritual sense similar to that of a medieval cathedral, creating its own light. Many visitors to the Arctic are from temperate zones where the land has been shaped by human hands, which is not the case in these high latitudes. Consequently, it takes time for temperate visitors to comprehend and appreciate Arctic vistas.

1 EXPLORING THE MARITIME FAR NORTHEAST

TEN THOUSAND YEARS OF ABORIGINAL settlement and five hundred years of European exploration and occupation have created connections and a common heritage among the lands and peoples from Maine to Greenland. What is known of this history comes from research into the geography, biology, and ecology of these lands; from archaeology of ancient Indian, Inuit, and Viking settlements; and from reports and records of early European explorers, fishermen, trappers, traders, and missionaries. Most of these European ventures were fleeting, and even less is known about prehistory. As a result, today these lands are still little known and are rarely in the public eye. Not until satellite images became available did people even begin to get a realistic sense of the Arctic in relation to the rest of the known world.

There are many reasons for the persistence of this state of ignorance. Much of the land from Maine to Greenland consists of impenetrable spruce forest and bog, tundra barrens, or glaciated mountains, while its seas are stormy, seasonally frozen, or forbid navigation altogether during the winter. These territories are among the least-populated areas on earth and were the last, except for Hawaii and Antarctica, to be discovered and settled by humans. Known originally to Europeans from the northern voyages of the ancient Greek mariner, Pytheas, the medieval concept "*ultima thule*" (borders of the known world), and from Norse myth and sagas, this region has long resided at the fringe of history—but not today. Climate change is transforming these formerly ice-locked lands and waters from a frontier to a global crossroads. As glaciers retreat and Arctic sea ice thins, perhaps even disappearing completely

13

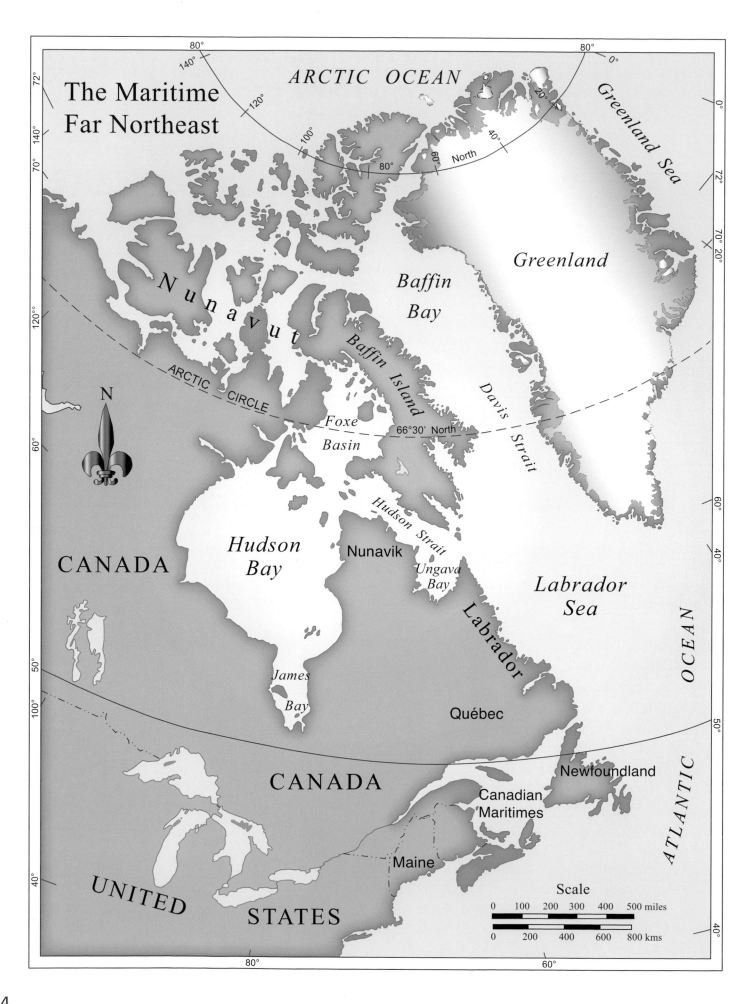

The Maritime Far Northeast

ARCTIC OCEAN

Greenland Sea

Greenland

Baffin Bay

N u n a v u t

Baffin Island

ARCTIC CIRCLE

Foxe Basin

66°30' North

Davis Strait

N

CANADA

Hudson Bay

Nunavik

Hudson Strait

Ungava Bay

Labrador

Labrador Sea

James Bay

Québec

OCEAN

CANADA

Newfoundland

Canadian Maritimes

ATLANTIC

Maine

UNITED STATES

North

Scale

| 0 | 100 | 200 | 300 | 400 | 500 miles |

| 0 | 200 | 400 | 600 | 800 kms |

1.03 Forest fire, Labrador
Because of Labrador's vast distances, limited transportation system, and low population of little more than 26,000, many fires—fueled by highly combustible caribou moss (*Cladonia rangiferina*) and impenetrable stands of spruce—are simply allowed to run their course.

< < < OPPOSITE

1.02 The Maritime Far Northeast
Spanning 40 degrees of latitude, from Maine in the south to Greenland in the northeast and Nunavut in the northwest, the Maritime Far Northeast has been connected by climate, geography, and animal life since time immemorial, and for the past 10,000 years by humans who have traversed its waterways for subsistence, trade, and industry.

during summers in a few decades, the world's nations are turning envious eyes to the North's thawing mineral treasures and its prospective Arctic sea routes. Climate change is making these lands and their resources more accessible and attractive to the outside world (figs. 2.03, 2.07–08). As this transformation begins, we need to understand these lands and their peoples and cultures as never before.

Today these northern lands and peoples are no longer as remote to southerners as they have been in the past. But the political and economic fragmentation that came with the rise of fixed national borders and trade barriers in the late nineteenth and early twentieth centuries has obscured some of the continuities in this macroregion. For the first time indigenous peoples who have lived there for generations, even millennia, are having more to say about how northern lands and resources are developed. Greenland is moving toward independence from Denmark and, in 1999, the Territory of Nunavut was established, giving Inuit peoples hegemony in the eastern part of Canada's Northwest Territories.

THE MARITIME FAR NORTHEAST

Because the Maritime Far Northeast (MFNE) extends through forty degrees of latitude and three climate zones—Temperate, Subarctic, and Arctic—and contains five of the Earth's ecosystems—sea, deciduous/mixed forest, conifer forest, tundra, polar/high mountains—it is a macroregion. Its constituent components, especially the ocean littoral, are shared from Maine to northern Greenland. In addition to being its own macrogeographic entity, it constitutes the northwestern portion of the North Atlantic Crescent, a term that refers to the lands arcing from Norway and Scotland to Newfoundland that all border a vast cold-water ecosystem. The concept of the "Far Northeast," developed by anthropologists (Strong 1930; Tuck 1975; Robinson 2006) in the 1970s to interpret the cultural history of northeastern North America became the organizing principle of this book, with a new emphasis on the maritime aspect. Despite its many political borders and natural boundaries, these lands forming the western shores of the North Atlantic have always been porous, interlinked by many waterways, common species of animal life, and shared cultural and linguistic ties. Viewed from a plane 30,000 feet above, similarities in the glaciated landscapes, drowned coasts, and sparsely vegetated and lightly populated lands are evident. These lands and waters, once connected by ice, are still linked by bird flyways and sea routes of myriad species of seals

1.04 The Farthest Far Northeast
Uummannaq, north of Disko Bay on Greenland's west coast, is inspiring. Explorers like Robert Peary, artists like Rockwell Kent, scientists, and explorers have all marveled at the unworldly landscapes of the northwest Greenland coast with their glaciated landforms, huge icebergs, misty vistas, and harsh rock.

1.05 Icebergs, Magical Shape-Shifters
Born of snow that fell on Greenland's ice cap as early as hundreds of thousands of years ago, when released from a glacier's grip may take as long as two years to drift south in the East Baffin and Labrador Currents. Only the largest reach the open Atlantic southeast of Newfoundland, one of which sank the *Titanic*.

and whales, by the migration patterns of codfish and caribou, and by the habitat requirements of creatures as large as polar bears and as small as lichens (fig. 1.06).

Cold currents provided the source of moisture that fed the growth of Ice Age glaciers that spread over these lands, lowered sea levels 300 feet, and extended to the limits of the northern continental shelves. Over eons, as ice ages have come and gone, glaciers carved the mountains and valleys into the form we know today. Geologically the region shares a common history of orogeny, the formation of mountains that created the northern Appalachian chain 480–380 million years ago. These chains include the Chic Choc Mountains of Quebec's Gaspé Peninsula and the Long Range Mountains of Newfoundland. A much earlier orogeny created the Torngat Mountains of Labrador and their extensions on the east and west sides of Baffin Bay.

During the Pleistocene—the most recent ice age, ca. 2.5 million to 10,000 years ago—the Laurentide Ice Sheet covered much of this coast, grinding down the highlands, carving U-shaped glaciated valleys, and dumping eroded sediments onto the continental shelf. The mountain glaciers of northern Labrador, Baffin, and Ellesmere and the ice caps of Ellesmere and

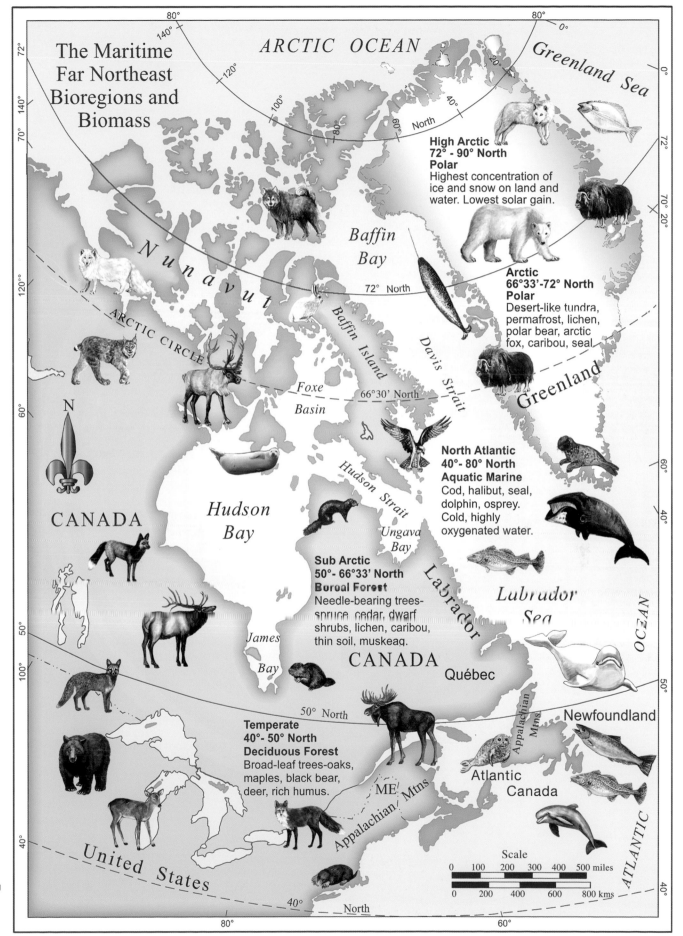

The Maritime Far Northeast Bioregions and Biomass

ARCTIC OCEAN

Greenland Sea

Baffin Bay

Nunavut

Davis Strait

**High Arctic
72° - 90° North
Polar**
Highest concentration of ice and snow on land and water. Lowest solar gain.

**Arctic
66°33'-72° North
Polar**
Desert-like tundra, permafrost, lichen, polar bear, arctic fox, caribou, seal.

Greenland

ARCTIC CIRCLE

72° North

Foxe Basin

Baffin Island

66°30' North

**North Atlantic
40°- 80° North
Aquatic Marine**
Cod, halibut, seal, dolphin, osprey. Cold, highly oxygenated water.

N

CANADA

Hudson Bay

Hudson Strait

Ungava Bay

**Sub Arctic
50°- 66°33' North
Boreal Forest**
Needle-bearing trees-spruce, cedar, dwarf shrubs, lichen, caribou, thin soil, muskeag.

Labrador Sea

Labrador

James Bay

CANADA

Québec

50° North

Newfoundland

**Temperate
40°- 50° North
Deciduous Forest**
Broad-leaf trees-oaks, maples, black bear, deer, rich humus.

ME

Appalachian Mtns

Atlantic Canada

United States

40° North

Appalachian Mtns

OCEAN

ATLANTIC

Scale

| 0 | 100 | 200 | 300 | 400 | 500 miles |

| 0 | 200 | 400 | 600 | 800 kms |

1.06 Bioregions and Biomass
This map, which utilizes latitude as a zonal scheme, addresses the carrying capacity of flora and fauna, land and water, as a function of "northernness."

1.07 Muskox

Muskox (*Ovibos moschatus*), native to Arctic regions of Canada and Greenland, have their natural habitat in northernmost Greenland. They are tolerant of winter moisture and have warm underfur called qiviut, prized by weavers. When attacked they form defensive rings, horns out, with young in the middle—a good defense against polar bears and wolves, but not against humans with rifles. Muskox have been transplanted to southern regions of West Greenland, where they have thrived.

1.08 Caribou

Caribou (*Rangifer tarandus*), photographed here in Labrador are the animal no preindustrial Arctic resident could live without. It is prized for its tasty meat, its back sinews used as thread by seamstresses since the time of Neanderthals, its antlers that provide tool-making material, and its hollow-core hair and thick pelt that can warm in the coldest of temperature. Their populations are, however, subject to major fluctuations that can cause havoc for humans depending on them exclusively.

1.09 Ringed Seal on Baffin Bay
A young ringed seal (*Pusa hispida*) became so disoriented that it could not find its breathing hole to return to safety under the ice. Given the size and sharpness of seal claws, we decided to drag this animal gently by its tail flipper to a breathing hole.

1.10 Bottlenose Dolphin
The bottlenose dolphin (*Tursiops truncata*) is common in the Subarctic, where it feeds on fish and occasionally on birds. A sociable animal, dolphins travel in groups that facilitate coordinated feeding behavior. Their habit of leaping and playing in the bow-wave of boats makes them the delight of northern mariners.

Greenland are survivors of this era, as well as the rocky islands, headlands, and deeply indented bays of granitic and gneissic rock that characterize these coasts. Throughout this region the surface geology presents a chaos of glacial features known as drumlins, eskers, moraines, glacial till, and outwash plains. Glacial erratics (fig. 1.21) often protrude from the thin soils of farmlands in Maine and New Brunswick and are dominant features of the boreal lands from Gaspé to Greenland. As the continental ice sheet melted away from the coast, a corridor opened up between the retreating ice and the sea to create a route for plants, game animals, and early Indian peoples to migrate into the Maritime Far Northeast.

The physical conditions of this region's geography establish natural parameters for life (fig. 1.06). The shape of the continents, which determines how land interacts with the southward flow of cold Arctic water in the East Greenland and Labrador currents, is the most important of these conditions determining the region's climate and ecology. The coasts of the Maritime Far Northeast are bathed in these cold waters with chilling effects that extend as far as southern Maine and create the physical conditions for the biological congruence of species and shared human history.

The cold, nutrient-rich waters flowing from the Arctic provide physical and biological unity to this macroregion (fig. 2.09). Although ice-covered much of the year from northern Greenland to Newfoundland, the waters of this region vary in temperature only about 10° C (18°F), compared with a range many times larger for the region's land surface temperatures. As a result the marine system supports a far greater variety of species than is present on its more climatically harsh interior. The Maritime Far Northeast produces much of its biomass in its seas and forests. The cold, highly oxygenated waters of the North Atlantic produce a massive quantity of phytoplankton and zooplankton. From these minute life forms emerge a highly complex network of trophic conversions, which accumulates as a rich yield of biomass. The productivity of this maritime ecosystem is conditioned by western winds and ocean upwellings and by the North Atlantic Drift, known popularly as the Gulf Stream.

Most important for humans are its sea mammals, which range widely throughout the northwestern Atlantic (figs. 1.09, 1.10, 1.12). Seals, porpoises, and whales historically have supplied northern indigenous peoples with blubber for fuel, skins for clothing and boat covers, sinew for sewing and line, bone and ivory for tools, and food. Anadromous fish—especially salmon, trout, and char—not only support coastal communities, but their summer spawning migrations also make them seasonally available deep into the often food-starved interior.

Biologically the MFNE shares many species of plants and animals. Inland from the southern coast are mixed northern temperate forests of pine, maple, and birch. The region's forests produce biomass as a function of the concentration of its deciduous and coniferous trees and surface area of leaves and needles, which engage in photosynthesis. As one proceeds north the volume and variety of biomass are reduced. Deciduous forests are replaced by thick boreal forests (fig. 1.11), which thin and become more open with the addition of mosses, lichens, sedges, muskeg bogs, wetlands, and an occasional stand of boreal trees such as spruce, fir, and larch. The boreal life forms, which began establishing themselves 12,000 years ago as glacial ice sheets withdrew and eventually melted, are the dominant ecosystem; a great band of pine and spruce stretches from Alaska and Western Canada to Quebec, Newfoundland, and Labrador. The northern parts of the region are clad in tundra with their diminutive flora of sedges, lichen, mosses, and willows.

Each of these ecosystems varies in its ability to support life. The greater the concentration of energy in the form of sunlight that falls at a given location, the greater the amount of biomass produced to support the web of life, from phytoplankton and zooplankton to *Homo sapiens* and—in some locations in the Maritime Far Northeast—to the polar bear, *Ursus maritimus*. The forests create habitat for deer, moose, and woodland caribou; grizzly (extinct since 1850 in the east) and black bear; wolves and a host of smaller fur-bearers including fox, lynx, wolverine, mink, and many others. Tundra and polar or high mountain ecosystems with myriad dwarf plants are limited in their ability to synthesize light into chemical energy, but still support caribou, muskox, wolves, foxes, and a few other species.

Cultures and history are no less interconnected. Today anthropologists have determined that the Innu and Inuit have separate historical roots—the former from the northeastern forests and coasts and the latter from Asia via the Bering Strait, Alaska, and the Canadian Arctic. Modern Inuit culture can be traced back to the thirteenth- to sixteenth-century Thule culture of the Eastern Arctic and from there to Alaska. Inuit have lived in these Arctic lands for more than 4,000 years, and Maritime Archaic Indians occupied the western North Atlantic coasts from northern Labrador to Maine thousands of years earlier. An archaeological trail of their spear points, made of translucent Ramah chert from Ramah Bay in northernmost Labrador, can

1.11 Labrador's Boreal Forest
Labrador has some of the most productive boreal forest in eastern Canada. This view looks west up Grand Lake, about fifty kilometers north of Goose Bay. These forests supported a major fur trapping industry from 1750 to 1950 before fur farms and market collapse reduced their value.

1.12 Polar Bear, Northern Labrador
Arctic ice drifting south in the Labrador Current often carries seal-hunting polar bears as passengers. Bears on ice even reach northern Newfoundland. Today, with higher ocean temperatures, the ice pack has disappeared along the Labrador coast, and polar bears must survive on dry land during the ice-free months, which grow longer each year.

be traced from Hudson Strait to Maine, with sporadic finds even further south. After 1950, the advent of radiocarbon dating provided evidence for early Inuit occupations of the Eastern Arctic and Greenland beginning as early as 4,500 years ago, while it was established that Paleo-Indian cultures in the Gulf of St. Lawrence had pioneered habitation of the continental shelf around the Gulf of St. Lawrence 11,000 years ago. As ice retreated from the maritime shelf and coastal lowlands into the Labrador-Quebec interior, these early Indian groups pushed north into Labrador and remained its sole occupants until Inuit ancestors arrived from Alaska via the Canadian Arctic about 4,000 years ago (see Chapter 5).

While both Arctic Inuit and northeastern forest-dwelling Indians had origins elsewhere, they met and for the past 4,000 years have shared boundaries in Quebec, Labrador, and Newfoundland. Living side-by-side throughout this northeastern region, with Inuit/Eskimos to the north and Indians to the south, over time these groups shifted north or south of their current coastal borders in Labrador and Newfoundland, replacing one another in response to changes in climate and environment, including forest-tundra boundaries and sea ice conditions (Fitzhugh and Lamb 1985). From then until the present, the Indian cultures that occupied the lands from Maine to northern Labrador have displayed continuities in housing, technology, and ways of living that distinguish them from Indian cultures to the south and west, giving them a distinctive character. The same is true for the Inuit people and their Paleo-Eskimo ancestors. Virtually the same Eskimo-related cultures have been found from northern Greenland to southern Labrador and Newfoundland.

The concept of a Maritime Far Northeast emerged as anthropologists and archaeologists expanded research north from Maine and the Canadian Maritime Provinces into Newfoundland and Labrador. Junius Bird and Duncan Strong, pioneers of Labrador archaeology, conducted excavations and ethnological work in Labrador in the late 1920s and 1930s. Strong's "Old Stone

21

1.13 Lichen Assemblage
This hoary lichen with glacier-like growth patterns has accommodated a newcomer in its midst.

1.14 Purple Saxifrage, Nunavut
Adams Island, a tiny dot of land located on the south edge of the Northwest Passage, is largely abiotic—with the exception of lichens and the tiny purple saxifrage (*Saxifraga oppositifolia*), the most brilliant flora of the Arctic.

Culture" of Labrador pushed archaeological discoveries back thousands of years and proposed that Eskimo culture originated from early maritime Indian cultures of the Northeast. Strong's theory was later discredited when the origins of Eskimo cultures were discovered around the Bering Strait; but his idea that early Indians had lived like Inuit in Labrador thousands of years before the Inuit arrived was a prescient and long-lasting contribution toward the concept of Indian and Eskimo continuities in the Maritime Far Northeast. New impetus for the concept of a unified culture-historical zone came with James Tuck's proposal of a "Maritime Archaic" cultural continuum for coastally adapted Indian cultures stretching from northern Labrador to Newfoundland and Maine 8,000–3,500 years ago (Tuck 1975; see also Fitzhugh 2006; Robinson 2006; Bourque 2012). Besides the similarities in their maritime economies, these early cultures shared an elaborate burial complex, similar types of tools and art, an extensive coastal trade network, and large oceangoing canoes (see Chapter 5).

Economically, the peoples of the Maritime Far Northeast have been variously connected through fishing, whaling, sealing, and forestry. The earliest Maritime Archaic Indians developed from Paleo-Indian big-game hunters whose economies shifted from mammoths to caribou, walrus, and seals as the St. Lawrence emerged from the Ice Age and began to flood when glacial ice melted and the seas rose 12,000 years ago. Maritime Archaic Indians lived along these coasts from 9,000 to 4,000 years ago, harvesting seals, walrus, caribou, salmon, and codfish from Nova Scotia to northern Labrador; from Newfoundland to Maine they hunted seals and swordfish. The first Eskimos and their Inuit descendants arrived from the North when early Holocene warm climates were replaced by cooling conditions and expanding sea ice about 4,000 years ago. The European discovery of northwestern Atlantic lands and marine resources in the early 1500s brought a cascade of whalers and fishermen beginning with Basques, then Dutch, followed by Scots and Yankees from the sixteenth to twentieth centuries. During the ensuing five hundred years increasingly efficient exploitation decimated the whales and walrus. In recent decades that most precious commodity, codfish (*Gadus morhua*), the world's most important food fish, were driven nearly to extinction. After receiving international protection, most of the whales except for the North Atlantic right whale rebounded, but the Canadian population of codfish, which crashed in the 1970s, is only now beginning to climb back. Cod stocks on the United States side of the border did not crash and so a fishing ban was not imposed on the Gulf of Maine and

Georges Bank. The restrictions on harvests have not been sufficient for cod stocks to return to sustainable levels: cod stocks today are estimated at 18 percent of what is needed to maintain a healthy fishery in the Gulf of Maine and 8 percent in Georges Bank. In 2013 highly restrictive quotas were imposed for the first time on cod fishing (Seelye and Bidgood 2013).

Equally important are the region's terrestrial resources: timber, animals, plants, and minerals. For almost 10,000 years a variety of Indian cultures living in the forested coasts south of northern Labrador harvested timber for dugouts and birch bark for canoes, containers, and burial shrouds. Animals provided meat, skins, sinew, fur, and feathers for clothing and adornment; berries and plants were used for food, dyes, and fiber. From AD 1000, when Vikings settled in treeless Iceland and Greenland, they ventured from these homelands to the Gulf of St. Lawrence, describing and naming lands with reference to their resources: Markland or "forest land" (Labrador), and Vinland or "grape land" (Newfoundland and the Gulf of St. Lawrence). While the Norse were primarily interested in timber and grazing lands for livestock, they also needed grapes, soapstone, iron, and hardwoods for tools, weapons, and implements. These resources were more easily accessible to the Greenland Norse on the western shores of the Atlantic (fig. 1.18), which were closer than Europe. Norse explorations by Bjarney Herjolfson, Leif Erikson, Thorfinn Karlsefni, and others ca. AD 1000 lasted only a decade or two but provided the first documentation by Europeans. Norse sagas describe the lands and resources and tell of numerous encounters with Natives they called "Skraelings." Despite changes in vegetation from the rocky coasts of Helluland to the grassy shores of Vinland, the conditions of the coast and its marine resources were similar throughout, guaranteeing that Norse settlements could be sustainable and possibly profitable—had they not already been occupied and defended by Native people.

Fifty years after the Norse abandoned their

1.16 Earth, from an Earlier Age
A glacier gives up its potpourri of rock and sediment on northern Ellesmere Island. This earth is emerging from the last ice age—the Pleistocene, 2.3 million years in duration—and is still devoid of the simplest life forms. Even lichens, the first form of life that colonize environments exiting from glaciation, are missing. The moraine of black rock is about the height of a three-story building.

1.17 Torngat Mountains, Labrador
Just before entering Ramah Bay
from the south, a panoramic view of
the Torngat Mountains awaits the
traveler. The term "Torngat" represents the master spirit of seals, fish,
and especially caribou.

Greenland colonies around 1450, Europeans returned, this time to the southern portion of the region when Basque, French, and English voyagers began exploiting its fisheries, especially codfish, walrus, and whales, and later its furs. Minerals and furs became important targets after 1500. Martin Frobisher's expeditions of 1576–78 were financed by the false promise of gold. Following detailed mapping by Dutch whalers in Labrador and Greenland, and by Britain's James Cook in Newfoundland, everything but the central Canadian Arctic was reasonably well known to European navigators and commercial interests. During the heyday of European exploration in the mid-nineteenth century, even the central Arctic and the fabled Northwest Passage were charted and basic understandings of these inner recesses of the MFNE became known.

While these voyages connected Europe and the northwest Atlantic coasts, they created few lasting linkages within the region beyond those that already existed among its Native cultures. Basque and Dutch whalers in Newfoundland and Labrador in the sixteenth and seventeenth centuries conducted incidental trade with Mi'Kmaq, Innu, and Inuit peoples but established few long-term associations (see Chapter 5). Isolation characterized the communities of Europeans and Native peoples in this region until the nineteenth century, when whalers out of New England and Nova Scotia began venturing to Baffin Bay, where they competed with British and Scots; this process stimulated collaboration with Inuit who had been hunting whales there since the 1200s. When whale stocks declined, whalers were replaced by fishermen who exploited the region's bountiful codfish, seals, and walrus. By the end of the nineteenth century southern vessels were plying the seacoasts from Maine to northern Baffin and Greenland.

CONGRUENCE OF GEOGRAPHY AND HISTORY

The belief that the world's northern regions were only marginally habitable for humans has been disproven by archaeological studies, but these views remain stubbornly persistent. These Eurocentric attitudes originated with the tragedies of early European exploration ventures into North America's Arctic regions. Although

1.18 L'Anse aux Meadows
This Viking Norse settlement, dating to ca. AD 1000, in northern Newfoundland, is the first—and so far only—authentic Viking site known in North America.

they were familiar with European shores of comparable latitude, those coasts were bathed in warm Gulf Stream (known scientifically as North Atlantic Drift) waters, which left early European explorers ill prepared for the icy seas and frozen lands of northern Greenland and the Canadian Arctic. Yet even as their vessels were trapped or crushed by the ice, and their crews starved or died of scurvy—which was the fate of the Franklin Expedition in the 1840s—these explorers found themselves surrounded by Inuit who were prosperous and healthy, if not especially numerous.

These derogatory ideas about the North were further developed in the mid-to-late nineteenth century, as science began to trump theological beliefs about the nature of the world. Geographic determinism posited that culture and history were preordained by geography and environment, and further, that the planet's temperate zones possessed the optimal mix of geography and environment. In the Maritime Far Northeast a drop in human population density is a correlate of increased latitude. In Maine, New Brunswick, and Newfoundland, population is measured in terms of tens of people per square kilometer or square mile.

In Labrador, this measurement decreases to tenths of people, and in Nunavut and Greenland to hundredths of people. Thinning population density is a direct result of the fact that the further north the location, the lesser the annual amount of solar gain. This reduction is caused by the increased oblique angle of the sun. The greater that angle, the greater the angle of incidence and the amount of atmosphere which the sun's rays must penetrate, resulting in a corresponding reduction in absorbed solar energy and biomass—both in number of species and in carrying capacity. Michael Webber (2009) refers to this phenomenon as the "energy density of light." It is with these data in mind, combined with the experience of actually walking in these lands, that one comes to appreciate both the immensity and limitations of northern regions and their implications for human culture.

In the nineteenth century Europeans as well as Americans continued to venture forth, exploring and taming the lands and subduing the indigenous peoples of North America, Australia, Africa, and elsewhere. The series of cultural revolutions that produced enhanced technology, agriculture, and industry has allowed

1.19 Spring Fishing in Greenland
Greenland Inuit fish though the land-fast ice when warmer weather arrives in spring. Snow machines have become popular, but some Inuit in Canada and Greenland still use dog teams.

1.20 Ice Pans in Cabot Strait
Traveling by ferry across Cabot Strait became problematic in early April 2003 when a 100 mile-wide raft of ice pans impeded the boat's course. Harp seals choose this type of ice for their birthing and whelping grounds. Unfortunately, for much of the last decade, sea ice has become a rarity, greatly compromising the birthing of harp seals.

humans to dramatically bend nature to our needs. With increasing powerful technologies, humans believed they had the power to continuously change the world to their liking, effectively banishing environment as a controlling factor. In the twenty-first century, we are beginning to question this arrogant approach as the deleterious effects of our wasteful policies have begun to emerge. After almost three centuries of industrialization, unintended consequences are becoming all too evident. The earth's climate is changing, at least partly due to human action. Just a few decades ago, who would have imagined the possibility of an Arctic Ocean entirely free of ice in summer by the mid-twenty-first century? Change is now occurring in so many ways and at such a rate that we may be reaching a tipping point with the same irreversible and disastrous consequences—extinction—we have visited upon many other species on the planet. It is imperative that we learn from the few subsistence-oriented cultures that remain to make more sustainable use of the world we have helped create.

In the second decade of the twenty-first century we are experiencing an increasing number of unintended consequences of our actions, born of the dominance of our technological culture and huge population. With seven billion of us now living on the planet, our continued and complacent reliance on technology threatens our existence. We are filling the atmosphere with carbon dioxide, carbon particles, mercury, and other chemicals, with a dramatic reduction in animal species, decimation of ocean life, and pollution of the aquatic food chain as results. Industrial agriculture has produced eutrophication, fouling essential habitats with chemical runoff and exacerbating soil loss in places like the Mississippi River Valley and Delta. Northern lands are no longer being protected from the pounding sea by ice, and the melting permafrost is turning Arctic lands into mosquito-infested marshes on which neither travel nor hunting is possible, while emitting huge volumes of methane.

POLAR EXPLORERS: PEARY AND MACMILLAN

By Susan A. Kaplan

1.22 Robert E. Peary
Peary made many geographical discoveries, but he remains a controversial figure because his North Pole claims could never be verified.

O N APRIL 6, 1909, ROBERT E. PEARY (1856–1920), his assistant Matthew Henson (1866–1955), and four Inughuit (Seeglo, Ootah, Ooqueah, and Egingwah) stood at the North Pole. For Peary and Henson it was time to rejoice, for this marked the culmination of years of grueling work. For the Inughuit (Inuit from Northwest Greenland) it was just another day of a strange and unsettling journey over dangerous, moving sea ice in an area devoid of game. Little did any of them know that one hundred years later the place where they had stood on that April day would be wrapped in controversy. Nor could they have imagined that in 2013 nations would be plotting shipping routes through the middle of the Arctic Ocean and vying to lay claim to the North Pole and its energy resources under the sea floor.

Peary, who explored the Arctic from 1886 to 1909, announced his North Pole achievement from southern Labrador in September 1909. Days before the world received his news, Frederick Cook, another American Polar explorer, made a similar claim from Denmark, reporting he had stood at the North Pole a year before Peary. An acrimonious debate ensued about the veracity of their respective claims, with neither man providing definitive proof of his achievements.

To this day people debate whether it was one of these men or a later explorer who first stood at the North Pole. The debate has reduced Peary, Cook, and Henson to one-dimensional figures. Meanwhile, history has largely ignored the contributions of the Inughuit, without whom neither party would have met with any success. Scholars are now building more complete pictures of the people

1.24 S.S. *Roosevelt*
Peary designed the *Roosevelt* and used it on his 1905 and 1908 North Pole expeditions. A forerunner of modern ice-breakers, its innovations included a sloping bow and powerful engines that could force a passage by breaking ice rather than skirting it.

involved in these expeditions and the social, economic, political, and scientific impacts of their endeavors.

An examination of Peary's original journals and correspondence reveals that this Bowdoin College graduate (class of 1877) was constantly improving the design of equipment used on his expeditions. Beginning in the late 1890s he experimented with various sledge styles, borrowing Inughuit designs and construction techniques. Sledging parties needed to drink hot beverages while on the sea ice, but Peary found commercial camp stoves unsatisfactory, so he designed a lightweight stove that turned ice to boiling water in seven minutes. His most ambitious project was conceiving the design of the SS *Roosevelt*, an auxiliary steam vessel that in 1905 and again in 1908 carried his crews to the northeastern shores of Ellesmere Island, where it then served as expedition headquarters throughout the fall, winter, and spring. The *Roosevelt*, a forerunner of the modern icebreaker, had the power and strength to push its way through the ice-choked narrow waterway between Ellesmere Island and Northwest Greenland, ramming the ice rather than dodging it. The *Roosevelt* was a product of Maine ingenuity: it was constructed in the McKay and Dix shipyard on Verona Island, and its powerful boilers were built by the Portland Company. The vessel's chief engineer, George Wardwell, was a Bucksport, Maine, man whose resourcefulness and mechanical skills were essential to the expedition's success.

1.23 Eagle Island
In 1904 Peary built a summer home on Eagle Island in Casco Bay, where he sought refuge from the North Pole controversy and began his involvement with airplanes. Eagle Island is now a state park open to the public.

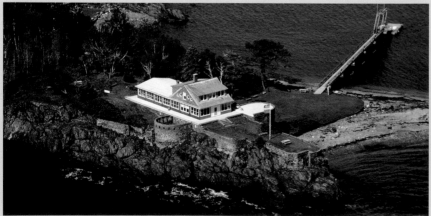

Matthew Henson was one of the most important members of Peary's team for more than twenty years, much respected by Peary's crews and beloved by the Inughuit for his knowledge, skills, kindness, and gentle manner. He was a gifted dog-sledge driver who learned to speak excellent Inuktitut. Although Peary credited Henson as invaluable to his conquest of the North Pole, as an African American, Henson could not attend the celebratory dinners held for expedition members when they returned to the United States, nor was he at the many ceremonies when other expedition members received accolades, medals, and awards. His achievements were finally recognized in the 1940s and 1950s when he received a silver medal from the US Congress similar to the one awarded to Peary decades earlier, and other organizations recognized his achievements and invaluable service to the nation.

Peary's Arctic expeditions were well known at Bowdoin, where another Bowdoin graduate and an avid outdoorsman, Donald MacMillan (1874–1970) became a member of Peary's 1908–09 crew. That winter MacMillan moved supplies over the sea ice by dog sledge, recorded tidal readings for months at a time, and hunted with Inughuit companions. His introduction to northern work was life-changing; he spent the next 45 years going north and died in 1970 at the age of 96.

In 1913 MacMillan led a large expedition to Northwest Greenland where he planned to lay claim to Crocker Land, a landmass sighted by Peary northwest of Axel Heiberg Island. The landmass turned out to be a mirage and the two-year expedition turned into a four-year venture when relief ships failed to retrieve MacMillan's party. He and the other scientists on the expedition made good use of their time, mapping glaciers, studying the floral and fauna, recording weather conditions, maintaining tide records, compiling Inuktitut word lists, and taking some of the earliest motion-picture film footage shot in the Arctic.

In 1920, with a number of expeditions under his belt, MacMillan decided that he needed to command his own ship. Borrowing design concepts from both the *Roosevelt* and the *Fram*, the vessel used by Fridtjof Nansen in his circumnavigation of the ice-bound Arctic Ocean in 1893, MacMillan contracted with the Hodgdon Brothers Shipyard of East Boothbay, Maine, to build a two-masted auxiliary schooner. The vessel, designed by William Hand, was 88 ft. long, 21 ft. wide, and weighed 66 tons. It was sheathed in ironwood and reinforced with concrete. The schooner had a spoon-shaped hull, so when the vessel was pinched by pans of ice it would slide up and rest on them rather than be crushed. Underway, this nimble and

1.27 *Bowdoin* in Battle Harbor, Labrador
Now a tourist destination, Battle Harbor was the station from which Peary sent a radio broadcast announcing that he had reached the North Pole.

sturdy vessel easily cut through ice. MacMillan named his schooner *Bowdoin*, after his alma mater; today, it is the official vessel of the State of Maine, sailed through North Atlantic waters by cadets at the Maine Maritime Academy in Castine.

Like Peary, MacMillan helped test new technologies that transformed the way people work and live in the Arctic. He began experimenting with Wireless radio transmissions in 1913 and finally succeeded in 1923. His Wireless North Pole station on the *Bowdoin* eliminated the radio silence that had enveloped previous northern expeditions. When *Bowdoin* spent the winter in Refuge Harbor, Northwest Greenland, he maintained two-way radio communication with people in distant parts of the globe.

In 1925 he and Richard Byrd co-led an expedition to Northwest Greenland, where US Navy pilots tested the flight capabilities of amphibious aircraft. In 1927 he outfitted a Model-T Ford with skis and used it for winter travel in Labrador. He worked for the United States during and after World War II, sharing his knowledge of the northern geography. MacMillan was involved in humanitarian efforts as well, providing residents of Labrador and Greenland with medical equipment, school supplies, and useful technologies.

Objects associated with Peary's Arctic ventures, including one of the five sledges that reached the North Pole, along with MacMillan's papers, photographs, films, and artifacts are housed at the Peary-MacMillan Arctic Museum on the Bowdoin College campus in Brunswick, Maine. These collections, documenting a history shared by residents of Maine and the Arctic, are a treasured legacy that helps introduce new generations to the wonders of the Arctic.

1.28 Ramah Chert Blank
A "blank" is a piece of Ramah chert that has not yet to be completely fashioned. Ramah Chert is translucent, as can be seen in this blank, found in the village of Kégaska on the Lower North Shore of Québec.

1.29 Cotton Grass in Nain
Cotton grass is abundant in Labrador wherever there is abundant fresh water, as here when a stream enters the bay. Inuit used cotton grass for the wicks in their oil lamps and as absorbant material for baby diapers.

Yet in these northern communities one is constantly aware of the strength of the social safety net.

Native peoples have learned that societies that give up their land and their knowledge of local subsistence strategies are doomed. Buying into the rest of the world's paradigm of growth and specialization is only a short-term survival option. Harsh experience has taught northern communities that overreliance on external opportunities is a recipe for disaster. The ecology of northern communities teaches us to be receptive to messages from the land and that economic diversity is the key to long-term survival.

Taken collectively, these northern cultures with their intimate connection to environment have fostered practices that encourage less use of fossil energy, a stronger sense of sharing and community life, reliance on family and multigenerational structures, greater use of sustainable local food production, and development of regional economic and political structures and alliances. These are values that have sustained the small communities of the North for thousands of years despite low population levels and vast distances between settlements. We would do well to emulate these values found in the societies and cultures of the Maritime Far Northeast as we approach the challenges that lie ahead in this rapidly globalizing world.

have sown large crops of family strife, drug use, and social dysfunction in many northern communities. For Northern peoples, population centralization, industrialization, political considerations, and social and medical problems are issues that increasingly affect the quality of life. George Wenzel, an anthropologist who has worked extensively among the Inuit of northern Baffin Island, observed that Inuit "are trying just as hard today to adapt as they did 500 or 900 years ago; the difficulty is that they are adapting not to the Arctic but to the Temperate Zone way of living" (1991: 34).

ONE PEOPLE, ONE LAND, ONE EARTH

Today, the Maritime Far Northeast is under increasing stress from the concurrent forces of global

1.30 Uummannaq
This cleft peak, which resembles the Bishop's Mitre, a landmark in northern Labrador, crowns Uummannaq, one of the largest settlements in northern Greenland. The solid winter sea ice seen here is becoming rare in northern Greenland.

warming, melting ice, and impacts generated by the outside world and its rush to extract oil, gas, and minerals from northern lands and waters. The prospect of a seasonally ice-free Arctic Ocean has introduced new opportunities for trans-Arctic shipping and communication. Now the issue is not whether it is profitable to open an Arctic mine or explore for oil in northern Baffin or the Greenland Sea, it is how to manage new industries and protect the fragile Arctic ecosystem and its biota from ecological catastrophes like blown-out wells or oil spills from tankers transiting the Arctic Ocean.

As global civilization teeters on the precipice of dramatic and possibly irreversible ecological and economic change, we need to seriously consider cultural models and traditional practices that improve the health of the planet. Some describe the Arctic as the planetary canary in the coal mine. The traditional peoples of the Canadian Maritimes, as well as the Innu, Inuit, and some Mainers, with their intimate knowledge of nature, provide time-tested ecological models for a sustainable Earth. For hundreds of years each of these cultures has adapted successfully to the harsh conditions of a northern climate. In our increasingly global community both subsistence and agroindustrial cultures are struggling

with many of the same issues. We need to become more familiar with practices that have sustained societies that have lived close to the land for generations—even millennia—emphasizing cooperation over competition, promoting a better balance between individual and community values, and providing more social support for all.

In the following pages we follow these and other themes that have emerged from exploring a region that has remained little known to much of the wider world. With its distinctive geography, biota, cultures, and history, this part of the world is going to be much more important in the next half century than it was when a period of cold climate and maximal extent of Arctic sea ice limited its involvement in the international economy. While scientists measure and try to accurately predict future global warming and politicians ponder its consequences, northern people provide one of the most accessible models for coping with climate change, based on detailed knowledge of how climate has changed and continues to change their lives. With the polar region experiencing unprecedented change, we are painfully reminded that our species must become more cognizant of our global responsibilities and become better stewards of the cultures, lands, and waters that sustain us.

2.01 Melting Iceberg
The fresh water cascading from this massive iceberg north of the Arctic Circle in Uummannaq Fjord fell as snow on the Greenland Ice Cap thousands of years ago.

CLIMATE CHALLENGE AND HUMAN ADAPTATION

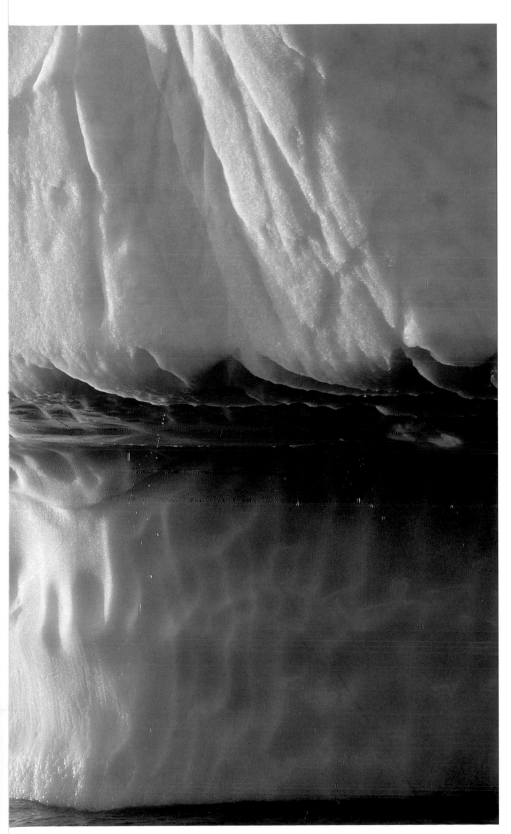

Winter temperatures plummeting six degrees Celsius and sudden droughts scorching farmland around the globe are not just the stuff of scary movies. Such striking climate jumps have happened before—sometimes within a matter of years… Most climate experts agree that we need not fear a full-fledged ice age in the coming decades. But sudden, dramatic climate changes have struck many times in the past, and they could happen again. In fact, they are probably inevitable…new evidence indicates that global warming should be more of a worry than ever: it could actually be pushing the earth's climate faster toward sudden shifts (Alley 2004: 62, 64).

IN 2007, FOR THE FIRST TIME IN RECORDED history, the Northwest Passage between Alaska and Greenland became ice free. Since then, Arctic sea ice has continued to shrink in terms of its thickness, extent, and longevity (fig. 2.07). It was reported that bowhead whales from the Pacific and Atlantic stocks met in the ice-free channels of the central Canadian Arctic when Arctic Ocean sea ice reached an historic summer minimum in 2007. In 2010 such a meeting was documented conclusively by satellite tracking of two tagged whales, one from the Pacific and another from the Atlantic, in Viscount Melville Sound in the Northwest Passage (Heide-Jørgensen et al. 2011). Genetic studies by these authors suggest that these whales must have been meeting sporadically during the past 500 years, which would account for the genetic similarities between the two stocks. In its 2012 "Arctic Report Card," the National Oceanographic and Atmospheric Administration announced a new minimum ice-cover record was established that exceeded that of 2007 by 18 percent, a huge reduction by any standard.

Throughout the Maritime Far Northeast, one single geological and climatological factor has molded the landscape: the Laurentide Ice Sheet, which repeatedly buried northeastern North America in ice to a maximum depth of 4.8 km (3 mi) over the past 2.3 million years. This ice sheet was about one mile thicker than Greenland's present ice cap and had more volume than the Antarctica Ice Sheet today. The Laurentide Ice Sheet was at least 50 percent larger than the combined ice masses in all of Europe and Asia (Pielou 1991). During the last full glacial period, the Late Glacial Maximum (LGM) of 18,000 years ago, the merged Laurentide and

2.05 Albedo Effect
Exposed dark rock soil or open sea as seen in this Greenland landscape absorb solar energy (low albedo) and warm the earth, while ice- and snow-covered lands and water reflect solar energy (high albedo) and cool the earth.

2.06 Arctic Warming Enhanced
Factors that enhance the effect of warming in Arctic regions are explained in this graphic.

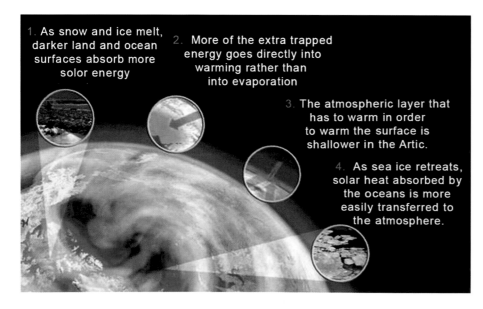

1. As snow and ice melt, darker land and ocean surfaces absorb more solor energy

2. More of the extra trapped energy goes directly into warming rather than into evaporation

3. The atmospheric layer that has to warm in order to warm the surface is shallower in the Artic.

4. As sea ice retreats, solar heat absorbed by the oceans is more easily transferred to the atmosphere.

different mechanisms, both bring heat from the southern latitudes to the North. The primary distinction between the two is that heat carried by the THCB is transported from the Pacific and Indian Oceans while the Gulf Stream flows from the much closer Gulf of Mexico and Caribbean.

Research on annually deposited snow layers in Greenland ice cores (Dansgaard et al. 1985; 1989; Alley 2002, 2004) shows that climate changes, including the onset and termination of major climate events, can occur very rapidly, on the order of decades or less.

> Global climate usually changes little over the course of a human lifetime, but a large and rapidly growing body of research has begun to reveal just how variable it is on longer time scales…transitions between fundamentally different climates can occur within only decades…[and there is] growing awareness of how profoundly human activity is affecting climate (Smith and Uppenbrink 2001: 657).

For the first time in decades, scholarly and public media have responded to these changes by running high-profile articles and programs featuring climate change, and the topic has become a central issue of national policy debates. NOAA and the National Science Foundation have devoted more resources to gathering climatological and geophysical field data, and scores of complex computer programs have been developed to model past climate change and future projections. Slowly, fields like the geosciences, climatology, and astronomy are contributing to better understanding of earth science systems (Lovelock 2006: 162). One of the convincing results is a strong relationship between the use of fossil fuels and climate change. Recently the term "tipping point" has been applied by ecologists,

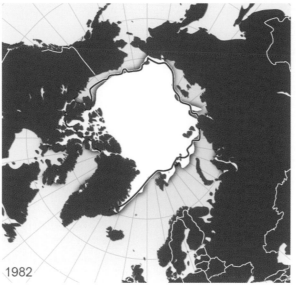

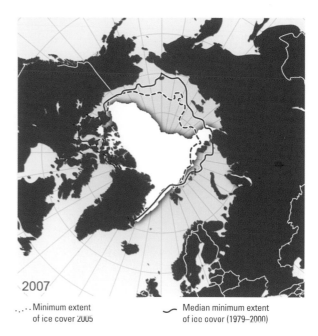

1982

2007

2.07 **Arctic Sea Ice Minima Mapped, 1982–2007**
Recent decades have seen a major reduction of summer sea ice cover in the Arctic Ocean, and predictions call for ice free summers by 2040 or sooner.

Sources: Fetters, F., and K. Knowles. 2002, updated 2004. Sea ice index, Boulder, CO: National Snow and Ice Data Center. Digital media. (ftp.//sidads.colorado.edu/DATASETS/NOAA/G02135/ Accessed October 2007)

.... Minimum extent of ice cover 2005

— Median minimum extent of ice cover (1979–2000)

geophysicists, and climatologists with reference to climate change. Global climate is at the point where a small increase in CO_2 may not just create a little change in temperature but lead to a dramatic change, for example, greatly increased glacial melting and storm activity. The amount of CO_2 measured in parts per million (ppm) has grown by about 40 percent from 270 ppm in 1750 to 389.78 ppm on 1 December 2011, according to researchers at Scripps Oceanographic Institution. In April 2012 NOAA reported a new milestone when for the first time a monthly average of 400 ppm was recorded at Barrow, Alaska. Although the cause of rising CO_2 has sometimes been contested, the data begin to reveal the hydrographic and atmospheric factors that force climate change. The real challenge now is to discover and then model the many feedback mechanisms that collectively constitute the climate matrix.

Sea currents, long investigated in general terms, are being studied more intensively as potentially one of the most powerful agents of climate phenomena (figs. 2.09–10). Discussions at the International Scientific Congress on Climate Change in 2009 revealed the lack of knowledge and disagreements among geophysicists, hydrologists, and other specialists about North Atlantic currents. H. P. Huntington et al. (2010: 265) assert, "neither local residents nor scientists have much detailed understanding of currents in any of the locations [of the Western Hemisphere Arctic] nor does anyone have an inexpensive way to monitor currents."

As the Arctic Ocean continues to change from a giant reflector of ultraviolet light to a massive sink of ultraviolet light, or as Arctic permafrost melts, releasing not only more CO_2 but also the much more potent greenhouse gas, methane (CH_4), from soils and seafloor

Observed sea ice September 1979

Observed sea ice September 2003

2.08 **Arctic Sea Ice Minima Observed, 1979 and 2003**

© NASA

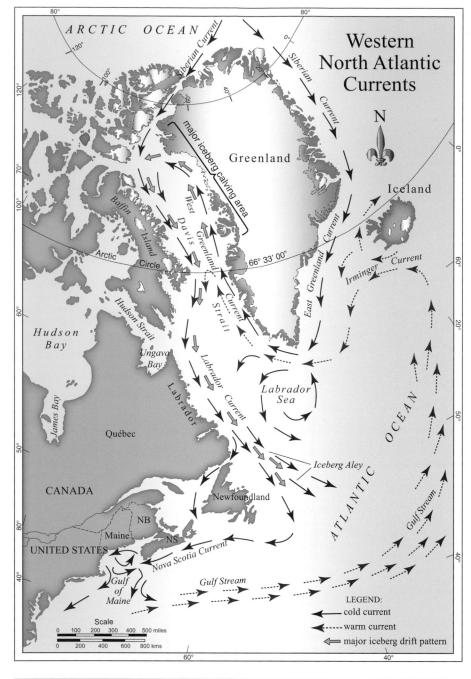

Western North Atlantic Currents

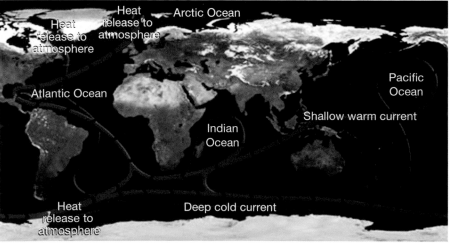

sediments, how much more will temperature rise? What are the implications for not only northern ecosystems but also for the whole planet?

Three incontrovertible facts have been established by late 2011 from the debate over the causes of recent climate change: sea levels are rising due to the net melting of the world's glaciers and volumetric expansion of seawater as it warms and expands; the production of CO_2 by humanity continues to increase across the globe; and human agency currently has greater impact on climate than either earthquakes or changes brought on by the earth's orbital ellipse (that is, the extent of tilt as the earth 'wobbles' on its axis). Paul Crutzen first (2002) used the term "Anthropocene" to emphasize the human origins of this era of dramatic warming on our planet; the idea has been popularized by Kolbert (2006) and Flannery (2005).

If the level of CO_2 in the atmosphere doubles, as predicted, by the end of the twenty-first century, we are likely to see a temperature increase of about 1°C (1.8°F) "if everything else is held constant…[but] everything else cannot be held constant" (Alley 2011: 85). Predictions assume a positive feedback loop; that is, more CO_2 will produce an increased warming effect. The effects of theses changes are difficult to measure in absolute terms, but they are contributing factors in observed changes in atmospheric water vapor, severity of storms, drought and flooding, changes in vegetation, and crop failures.

Even before the decline of sea ice began to make Arctic regions more accessible to outsiders, studies of atmospheric circulation showed that pollutants generated in the tropics and temperate zones are carried north by air currents, where they descend and are deposited in polar regions. The phenomenon of heat rising, transporting pollutants with it, can be observed when we sit around an open fire. Soot from coal-fired generating plants, aircraft engines, and other industrial processes is a major contributor to airborne pollution that reaches Arctic regions and hastens both global and Arctic warming. Soot and other carbon particulates also are deposited on snow on land and on sea ice, decreasing albedo and accelerating polar glacier and sea ice melting. During the summer, old multiyear sea ice in the Arctic Ocean becomes black with soot as the annual layers melt out, becoming blacker and more heat-absorbent as the melt progresses. Alun Anderson (2009: 249–252) notes that black carbon (ice dust) generated by industry, jets, forest fires, and annual spring burning by agriculture covers one-quarter to one-half of the Arctic between N50° and N60° latitude. Either

way, whether suspended in Arctic air or deposited on the earth or ice, soot captures solar radiation. In just twenty years, between 1990 and 2010, air temperature in the Arctic rose by 2.5°C (4.5°F).

A final factor contributing to Arctic warming results from the structure of the polar atmosphere: its troposphere—the lower portion of earth's atmosphere—is only half as thick as elsewhere and has a correspondingly diminished carrying capacity. Polar regions therefore have less capacity to absorb pollutants, which are flushed out and deposited in the Arctic. As a result polar areas contain higher relative concentrations of synthetic chemicals such as PCBs. Multiple studies of the marine food web reveal how organic mercury, a metal that is contained within airborne pollution from burning coal, builds up (Gray 2002). When these pollutants enter the food chain, they are biomagnified as they progress from phytoplankton to zooplankton, and to fish, seals, whales, and eventually to humans.

Change in much of the Maritime Far Northeast is occurring at an increasingly rapid rate. Stress factors—that is, a constellation of climate, environmental, and economic changes that are outside the normal boundaries of traditional cultural adaptations and ecological knowledge—are now being experienced by Arctic peoples. The implications of this "regime shift" or "rapid systemic change" have yet to be fully appreciated (Yalowitz et al. 2008: 5).

Climate change as it is currently being experienced can be traced to humans whose presence—although brief—on this planet exceeds the impact of any other species during Earth's history. Richard Alley emphatically lays the blame for global warming on our species, noting that while the rise in temperature from the Ice Age took "10,000 years, we may achieve a similar warming in not much more than a century" (2011: 186). This same mantra of human culpability is noted by others: "Continued loading of carbon dioxide into the atmosphere only increases the uncertainty and the instability with which we will have to contend. [Humans] have

TRANSIT OF THE PETERMANN ICE ISLAND

By Wilfred E. Richard

ALAN HUBBARD, A GLACIOLOGIST FROM Aberystwyth University in Wales, had been working at the head of Uummannaq Fjord in Greenland, measuring glacial movement, when we met in spring 2010. In 2009 and earlier in 2010, Dr. Hubbard had placed recording instruments on the Petermann and Humboldt glaciers, which are located in far northwestern Greenland on Nares Strait. These glaciers consist of both a land component and a floating ice shelf—the latter, an uncommon feature of Arctic glaciers. His instruments measured significant ice movement, evidence that the structural integrity of the Petermann ice shelf was rapidly degrading.

In August 2010, a 12-mile-wide, 3,000-foot-thick iceberg calved from the Petermann Glacier. This was the largest piece of floating ice released in Greenland in almost half a century. This floating massif, transported by the southbound Labrador Current, began its yearlong journey to Newfoundland and the Gulf of St. Lawrence.

When the Arctic Studies Center archaeology crew arrived in Newfoundland for our tenth season on Quebec's Lower North Shore in July 2011, we soon heard that a 10-mile-long iceberg had grounded in southern Labrador off the settlement of Battle Harbor. Within fifteen minutes of setting sail from Long Island, we encountered our first icebergs at the base of Notre Dame Bay. Usually we see one or two icebergs near the northern tip of Newfoundland, but in 2011 there were many icebergs all along the east side of Newfoundland's Northern Peninsula, in the Strait of Belle Isle, and in the Gulf of St. Lawrence. They had all calved from the stalled mother remnant of the Petermann Glacier.

In mid-August, on the return leg back to Newfoundland, local fishermen at Quirpon informed us that the Quirpon Tickle (narrow passage) was jammed with ice and would be very difficult to navigate through. After waiting for fog to lift, Skipper Perry Colbourne was able to weave through a plethora of massive icebergs, which were the prominent feature of the seascape until well into Notre Dame Bay. Later we learned that the mother iceberg located off Labrador earlier in the summer had broken free as two icebergs, which measured 2.8 x 3.4 miles and 2.4 x 2.2 miles. The *Pitsiulak* had motored through this concentration of floating ice off the east coast of Newfoundland. This ice had floated from 81 degrees North Latitude (fig. 2.14), the Arctic waters of Greenland, to 50 degrees North Latitude, the High Temperate zone of Atlantic Canada.

I called our skipper, Perry Colbourne, in October 2011, in search of a lost piece of photo equipement, and learned that consequences of Petermann Glacier breakup were still being experienced in Newfoundland. A ferry crosses the "tickle," about 0.4 miles wide, between Pilley's Island and Long Island, where Perry lives. That day Perry had counted 59 icebergs in the waters between Long Island and Pilley's Island; earlier the tickle was blocked by one very large iceberg, and ferry service had been temporarily discontinued. The melting of the Greenland Ice Sheet may deliver more than these inconveniences and maritime hazards, as climate change in this part of the world begins to ripple throughout the Maritime Far Northeast.

2.14 Petermann Glacier Ice
In 2010–2011 a massive iceberg, so large it became known as Petermann Ice Island, flowed south from Greenland. In August 2011, several very large icebergs calved off, including these photographed from the town of St. Carol's on French Bay, Newfoundland.

become the biggest force in the climate system" (Conkling et al. 2011: 22–23).

A consensus on human agency and the increase in greenhouse gases has been reached between two of the most involved US agencies. Addressing global climate, the IPCC report concludes: "*There is evidence that some extremes have changed as a result of anthropogenic influences, including increases in atmospheric concentration of greenhouse gases.* It is *likely* [IPCC emphasis] (66 to 100% probability) that anthropogenic influences have led to warming of extreme daily minimum and maximum temperatures on a global scale" (IPCC 2011: 6).

The National Oceanographic and Atmospheric Administration (NOAA) annually focuses on the Arctic, issuing a "Report Card." Its December 2012 Report Card states:

An international team of scientists who monitor the rapid changes in the Earth's northern polar region say that the Arctic is entering a new state, one with warmer air and water temperatures, less

summer sea ice and snow cover, and a changed ocean chemistry. This shift is also causing changes in the region's life, both on land and in the sea, including less habitat for polar bears and walruses, but increased access to feeding areas for whales.

NOAA cites record-low winter snowfall and sea ice extent; record-setting summer melting season in which 97 percent of the outermost layer of the Greenland Ice Cap experienced some thawing; increased vegetation growing season; and record-high permafrost temperatures in Alaska; warmer sea surface temperatures and massive phytoplankton blooms in the Arctic Ocean. According to Martin Jeffries,

The record-low spring snow extent and record-low summer sea ice extent in 2012 exemplify a major source of the momentum for continuing change. As the sea ice and snow cover retreat, we're losing bright, highly reflective surfaces, and increasing the area of darker surfaces—both land and ocean—exposed to sunlight. This increases the capacity to store

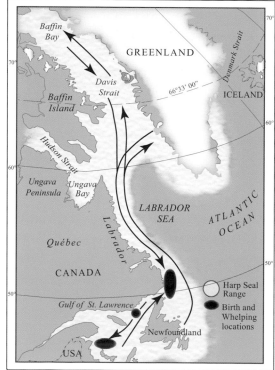

2.15 Harp Seals Basking off Labrador
Thousands of harp seals pause on their northward spring migration to Greenland and Baffin Bay, resting on the pack ice of the northern Labrador coast.

2.16 Harp Seal Migration Routes and Whelping Grounds

heat within the Arctic system, which enables more melting—a self-reinforcing cycle. (Jeffries, from NOAA 2012 Arctic Report Card)

How these changes affect northern life and cultures became evident to us as we pursued our travels and research in the Maritime Far Northeast.

PERSONAL OBSERVATIONS FROM THE FAR NORTHEAST

Personal experiences and reports of local observers have documented that the ice season in the eastern Arctic is now shorter and more erratic than in the past, and winds are higher and more frequent. As waters become increasingly ice-free, northern peoples are finding it more difficult to obtain food. Warmer water has enabled southern species of sharks to move north into Subarctic and Arctic waters, and orcas (killer whales) are now competing with Inuit and polar bears for seals. Insects, such as wasps, never seen previously in northern Greenland, have recently been reported in Uummannaq.

Particularly striking changes are occurring in harp seal habitats. Perry Colbourne, a resident of Lushes Bight, Newfoundland, and skipper of the Smithsonian Arctic Studies Center's archaeological research vessel, and Phil Vatcher, a former wildlife officer for the Quebec Lower North Shore (LNS), noted the decrease in the harp seal population in August 2010. Historically,

2.17 Polar Bear (*Ursus maritimus*)
This apex marine predator (other than humans) ranges throughout the Arctic. Its primary prey is the ringed seal, but it also takes sea birds and small walruses and, when on land, caribou. Polar bears that hunt on the southward-moving pack ice off Baffin Island and Labrador often end up in northern Newfoundland and are killed or darted for transport north when they come ashore and attempt to travel back north over land.

2.18 Greenland Sledge Team
The husky, originally bred as a sledge dog in Siberia, arrived in Alaska 1,500 years ago and helped make possible the Thule culture migration into the Eastern Canadian Arctic and Greenland around AD 1250. Teams of dogs can travel with a loaded sledge as much as one hundred miles in a single day.

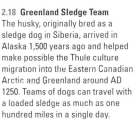

the harp seal has been a highly dependable seasonal resource for prehistoric and modern hunters in Labrador, Newfoundland, and the Quebec Lower North Shore (Stenson and Sjare 1997). The population of the annual fall and spring migrations has been estimated at 9–12 million animals in recent years. Harp seals require stable winter pack ice (fig 2.16) on which they give birth (fig 2.15) and tend and feed their pups from February until April (Lacoste and Stenson 2000; Lavigne and Kovacs 1988; Sargeant 1991). The only ice that appeared in the northern Gulf in the 2005–06 sealing season were thin ice pans that had drifted down from Labrador and quickly melted in the warmer waters around Newfoundland (Johnston et al. 2005). In 2009 and again in 2010, seals came as usual to the LNS, but the winters were so warm that little sea ice formed. In early winter of 2010–11 the absence of sea ice forced females to give birth on shore, and many of the pups, known as "white coats" because of their fluffy fur, were

2.19 Halibut Fishing on Spring Ice
Halibut and cod are important components of the West Greenland economy. This rank of halibut hooks will be baited and set as a long-line through a hole in the ice. Without stable ice winter fishing will not be possible for locals who lack large boats, and this subsistence resource for dogs and humans will be lost.

2.20 Old Arctic Hand
Raymond Buffitt , who worked for years as an economic development officer with the Department of Indian Affairs in the Canadian Arctic, explained the Inuit switch from dogs to snow machines from the 1960s to the 1980s.

2.22 Winter Ice Road >>>
For the past 800 years, from Northern Greenland to the Quebec Lower North Shore, the use of dog sledges for hunting, fishing, and winter travel has been the primary means of transport. Today it is almost absent outside of northern Greenland, where its use is declining as a result of climate and economic change.

abandoned by their mothers and starved or were taken by predators. Others, born on thin ice, drowned when the ice melted before the pups molted, forcing them into the water before their white coats were replaced by the adult seal hair required for swimming and feeding on their own. Hunters along the LNS estimate that thousands of young harps perished during the past two seasons as a direct result of warm winters and lack of stable sea ice. Scientific studies report similar harp seal declines throughout the Gulf of St. Lawrence and Newfoundland region and attribute them to warming climate (Johnston et al. 2007, 2012). Compounding the loss of young animals, adult harp seals have begun to abandon their winter range in the Gulf and become unavailable to local hunters. The resulting population decline will impact other regions of northern Canada and Greenland where harps are a major seasonal food for people and dogs and a source of income from sale of their valuable pelts on the European market. (The US Marine Mammal Act prohibits importation of these pelts or other sea mammal products into the United States.)

Warmer winters are also affecting plants. The bakeapple (*Rubus chaemamorus*), also known as cloudberry or salmonberry (fig. 2.21), a tasty fruit that grows in bogs, is a favorite seasonal food in Subarctic regions. Phil Vatcher, Lloyd Rowsell, and Gilles Mongait—all LNS residents—attributed the failure of the 2010 bakeapple harvest to a lack of snow, which caused the roots of the plants to freeze. Here and in Newfoundland, snow and sea ice were so scarce in 2010 that snowmobile travel on land and sea ice was not possible. The warmer weather in Newfoundland in 2010 also

brought reports of the arrival of new species, including the pine grosbeak and goldfinch.

As sea ice melts and melting permafrost turns the tundra into a bog, even local water sources may become undrinkable without filtration or sterilization. Some speculate that melting permafrost might release long-dormant bacteria like tetanus, for which, at least in Greenland, there is currently no vaccination program.

In Greenland the winters from 2008 to 2011 saw very little fast ice, the ice that freezes firmly to the shore and provides a road for dogsleds and snowmobiles. While some fast ice still forms, it is weakly anchored and easily broken up by storms and blown out to sea. Consequently seal hunting and fishing on the ice by dogsled (fig. 2.22) has become limited to a few weeks in mid-winter, and summer hunting by small boats has been restricted by the increased number of icebergs calving from melting glaciers. Food shortages are an increasing consequence of such limitations to hunting. Some families migrate to towns with the hope that it will be easier to sustain self and dependants, but this disrupts the familial traditions and patterns of life on the land.

Polar bears (*Ursus maritimus*), the best-known charismatic species of the Arctic, are experiencing stress caused by climate warming and reduced sea ice (Vongraven and Richardson 2011). Polar bears hunt by stalking seals on the pack ice. As the ice retreats northward into the Arctic Ocean in the summer, polar bears are forced to abandon their dwindling sea-ice hunting grounds and swim long distances to land. Although good short-distance swimmers, polar bears are not adapted for open-ocean swimming, and some—especially the young animals—drown before reaching

2.21 Bakeapple
The ubiquitous bakeapple (*Rubus chamaemorus*), or *chicoutai* in Innu, is the northern fruit of choice. It grows around the world in boreal latitudes in muskeg and wet boggy soils. According to local observers, bakeapple harvests are decreasing, perhaps as a result of global warming.

sledges, owners are forced to shoot the dogs they cannot afford to feed. The choice between dogs and machines is a real dilemma as neither is truly satisfactory. "[T]he line between nostalgia and progress is mixed, [in] that it must always—and can never can truly—be straddled, and that the danger and necessity of doing so is both palpable and real" (Haake 2002: 62–63).

Nuugaatsiaq in northern Greenland has historically enjoyed a long winter ice season, but in the spring of 2009 hunters could not safely use dogs as they risked falling through thin ice (fig 2.24). In the winter of 2009 and 2010, the waters of Uummannaq, the northern Greenland village, remained open, without ice, forcing hunters to hunt from boats (figs. 2.23, 2.25). As ice mass declines much of the food that has sustained Inuit life and culture for millennia is disappearing. Hunting is increasingly conducted by boat, but without sea ice, ring seal, harp seal, and polar bear populations will fall and be geographically displaced. Some fear polar bears could face extinction, while others wonder if a tipping point has already been reached in terms of the Arctic Ocean's albedo.

That tipping point portends a complex series of events, seen for example in the relationship between ice, seal, and hunters. Without sea ice as a hunting platform, the Inuit hunter, like the polar bear, must adopt new hunting methods, using boats rather than sledges or

shore. This has been a particular problem for polar bears in the Beaufort Sea, less so in the Eastern Arctic.

Melting sea ice also affects the human harvest of polar bears and other sea mammals, as hunters must travel further to hunt, causing them to replace dog teams with gas-guzzling snowmobiles or outboards. Owners incur debt to purchase or lease large boats that are costly to maintain. As motorized craft are displacing dog

2.27 & 2.28 Tank Farms in Uummannaq and Upernavik

In the heart of every village, town, settlement, or hamlet in Nunavut or Greenland, is a fossil fuel–tank farm. That energy is used to produce heat and electricity and to power boats, wheeled land vehicles, and snow machines.

heat homes, and support airplane, snowmobile, out-board, and ATV transport. The skyline of every Arctic village is now marked by at least one fuel-storage tank (figs. 2.27–28).

This industrial legacy substitutes liquid fossil energy for energy formerly obtained from the blubber of seal, walrus, and whale. Before the arrival of Euro-peans in the Arctic securing energy was conducted in Labrador, Nunavut, and Greenland by acquiring blub-ber as a by-product of hunting sea mammals for food. Compared with the global society's industrial supply system, Inuit methods have been small scale, and today seals continue to be hunted, with one or two hunters taking a few seals by rifle, harpoon, or net each year. Pelts are sold if the price is high but otherwise are used for making sealskin boots and other local craft products; all blubber and meat is consumed by people and dogs. Although rifles are an innovation, Inuit sealing constitutes a traditional practice as the hunt is embedded in Inuit kinship structure of sharing what is harvested from sea and land. In earlier days food and blubber also came from walrus and whale, and walrus also supplied ivory for tools and harpoon heads, and hides for covering the large umiaks used for whaling and long-distance summer travel. Walrus meat was used only for dog food. Whale meat was also used for dog food, and its inner skin, known as *maktaaq*, has

importing costly petroleum products. The shift from traditional gathering of carbohydrates directly from the land to that of employing imported hydrocarbons to leverage local carbohydrates has created an Arctic dependency upon imported oil, making this form of energy, "the lifeblood of modern Arctic settlements" (Dowdeswell and Hambrey 2002: 189). Modern life in the Arctic has seen an explosion in the need for imported energy to generate fuel for reserve stations, long been considered an Inuit delicacy , while its baleen was used for making baskets, buckets, snares, and fishing line. During the continuing controversy over the hunting of baby harp seals around Newfoundland and the subsequent U.S. ban on importing Canadian marine mammal products, the Western world has been "unable to recognize that Inuit subsistence was a matter of cultural right, as well as need" and that "food was only available through hunting" (Wenzel 1991: 55, 113). Much of today's developed world:

> has acted with little difference from earlier generations of southern imperialists. In essence, it has acted with same ethnocentric raison d'être that has characterized all but perhaps the earliest moments of Inuit contact with Western culture (1991: 180).

Heretofore, energy needs were met from one source: whether for heating, cooking, or lighting, the oil that fueled the lamp was locally rendered from sea mammal fat. Now, energy sources in much of the Arctic are not only imported, they are imported for specific applications such as gasoline for snow machines or diesel generators for electricity. In Maine, much of the heat for homes is derived from firewood that is locally gathered. Heating with wood allows us to transfer heat locally, reducing our need for imported oil-based energy. Fortunately, in both the Arctic and Maine local fuel resources are abundant (Fallows 2008). However, in recent years the Arctic has shifted from using local sources of energy to imported energy and consumer goods. Now most Inuit hunt seals by snowmobile. Raymond Buffitt (fig. 2.20) of Chevry, Quebec, who spent a career in the Arctic as an economic development officer

2.29 Fuel Barrel in Nunavut
The omnipresent rusting fuel barrel is the ubiquitous cultural hallmark of the twentieth-century age of fossil fuels. After thousands of years of using sustainable oil sources, petroleum has replaced the blubber of whales and seals for heating, lighting, and transport.

with the Canadian Ministry of Northern Resources, questions the belief that the Canadian government encouraged the Inuit to shift from dogs to snowmobiles. He recalled that when Inuit began resettling in villages in the late 1950s and early 1960s local resources no longer provided enough seals for dog food, and so the use of dogs had to be abandoned. In the central Canadian Arctic, snowmobiles were a later introduction that gave villages a wider resource zone to exploit. Dog-sledging has continued into the present in the northern parts of West Greenland because oil rendered from a seal cannot power a snowmobile, but blubber will fuel up a dog team.

The shift from dogs to snow machines has impacted the quality of local diets. Rising gasoline prices—around $1.50 Canadian per liter ($.91/quart) or about $6.00 Canadian per gallon in 2010—have made it extremely costly for Inuit to obtain wild game, known to many northerners as "country food." As foods such as fish and caribou are replaced by imported foods high in fat and sugar, the energy levels of the Inuit has dropped, and the shift has contributed to increased rates of diabetes and tooth decay (Kuhnlein and Receveur 2007).

The introduction of southern market strategies and use of fossil fuels by northern communities are beginning to subvert regional communities where the economies have previously been based on local ecological interdependence. Growing dependence on the global economy threatens to engulf them in same perils that endanger the rest of the world. As Safina notes (2011: 108), "the economy sits entirely within the ecology…Edward Abbey long ago observed that growth for the sake of continuous growth is the strategy of cancer." He warns of the inherent conflict of supporting environmental stewardship within a market economy: "The failure of markets to realistically price the destruction of living systems and the fuels we use to run civilization makes it economically attractive to risk the entire planet" (2011: 271).

As the world debates the issue of whether the world's climate is changing as a result of human activity and considers what mitigation measures might be applied, northern residents already have been experiencing a new climate regime for the past ten to twenty years. It is now recognized that the Arctic is a bellwether, the "canary in the coal mine." The presence or absence of sea ice—a condition reflecting simply the difference of one degree on a temperature scale—does make all the difference in the world. In the following chapters we will see the effects of these changes on societies and ecology in a series of northern environments from Maine to Greenland.

3.01 View from the Camden Hills
Located in Camden Hills State Park, Mt. Baty affords a panoramic evening view of Camden Harbor, with church spires and sailboat masts ringed by Maine's spectacular autumn foliage.

3 MAINE – THE SOUTHERN ANCHOR

[T] he poet, must from time to time, travel the logger's path and the Indian's trail, to drink at some new and bracing fountain of the Muses, far in the recesses of the wilderness… and 'not be civilized off the face of the earth.'

Henry David Thoreau, *The Maine Woods*

THE MAINE THAT THOREAU DESCRIBED in the nineteenth century persists in the twenty-first century—as an untamed place that bonds its people to its land and attracts visitors from America's urbanized east coast. Maine is the point of departure for all areas to the north within the geographical entity that we have identified as the Maritime Far Northeast, stretching to Greenland. Maine has also been a destination for tourism since the nineteenth century: in the national imagination, it remained a "rustic" destination, akin to the frontier, as a place where nature still was unrestrained. Maine's seasonal residents transformed the economy of the seacoast and to some extent its culture, building houses in a wide range of sizes and styles, but all euphemistically called "cottages"—in what had formerly been shipbuilding and fishing villages. "Vacationland" appeared for the first time on Maine license plates in 1936 and continues to be the primary identity for tourism in Maine.

The same forces of glaciation that shaped Greenland, Nunavut, and Atlantic Canada created Maine's geography, modifying its landscape and moving gigatons of ice as well as boulders, gravel, and sand into the Gulf of Maine. Glacial cycles over the past three million years scraped down its mountains and carved out its many river valleys, pushing sediments out to the edge of the continental shelf. The huge weight of the ice sheet depressed the earth's crust several hundred feet and lowered sea level 300 feet, approximately the depth of the continental shelf's outer limit. As the ice began melting 15,000 years ago, rising seas flooded over the depressed coastal landscapes, leaving marine terraces far inland along the river valleys. Then, relieved of the

55

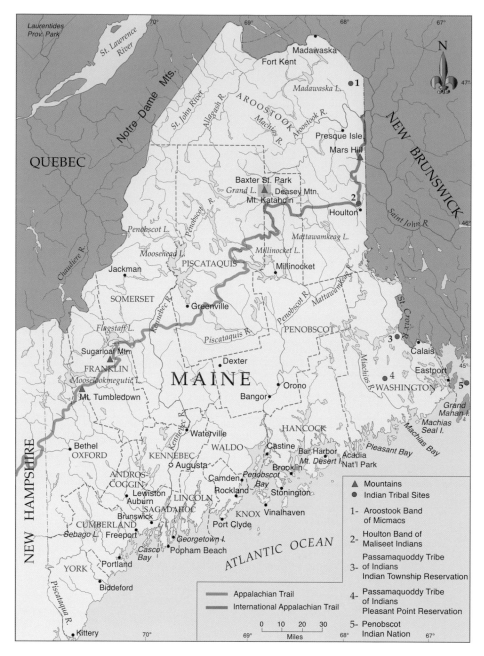

Most endangered are sites and properties in "soft-shore" oceanfront locations not protected by rocky cliffs or man-made barriers like sea walls.

The sea dominates Maine's eastern boundaries with a ragged coastline of 3,500 miles (5,630 km), the longest shoreline of any state on the Atlantic Ocean. Maine's many rivers carry food for shellfish and finfish from its mountains, valleys, and islands into the wetlands, estuaries, and waters of the Gulf. Rivers perform another function: as they flow into the sea, they displace seawater with fresh water, causing the nutrients to be mixed and more widely distributed (fig 3.03). Yet other factors that agitate the distribution of food throughout the Gulf are its storms and diurnal tides.

Geological forces have endowed Maine with its special geography. The drowning of its coast after sculpting by Ice Age glaciers and rivers created Maine's many ports and harbors, making it a vital entrepôt, first for the British colonies and later for the fledging United States. At the tidewater heads of Maine's largest rivers—Piscataqua, Presumpscot, Androscoggin, Kennebec, and Penobscot—port towns grew up and became manufacturing centers. The coast's parallel north-south rivers created many peninsulas known for their fishing ports and lighthouses.

The submerged coastal shelf is revealed in 3,166 islands, many rich in history. Islands were often named for foods they provided. There are fifteen Hog Islands, forty-two Ram Islands, thirty-four Sheep Islands, thirty-six Goose Islands, eighteen Duck Islands, seventeen Seal Islands, and eleven Egg Islands. Hogs, rams, sheep, and cows were often raised on islands to protect livestock from wildlife predation and to make fencing unnecessary.

The popular image of Maine is usually that of a coastal land; few outsiders realize that it has the largest forest in the east. More than 80 percent of Maine is forested, with its greatest tracts in the west and north. Maine has been blessed with both hard and soft timber and an extensive river system to transport logs. Its forests host an abundance of wildlife including many northern species of birds: the nonmigratory, hummingbird-size ruby-crowned kinglet (*Regulus calendula*: fig. 3.12), the restored and now populous wild turkey (*Meleagris gallopavo*: fig. 3.14), and occasional guests from the north, including the northern hawk owl (*Surnia ulula*: fig. 3.13).

POLITICAL GEOGRAPHY

The natural and political boundaries of the state reveal an affinity with its Canadian neighbors (fig 3.02). Maine is bordered to the west by New Hampshire, which occupies about 23 percent of its boundary;

ice burden, isostatic land-rise expelled the sea, and the shoreline dropped tens of feet below its current level, exposing a wide strip of land for animals and human settlement between 10,000 and 5,000 years ago. As the final stages of glacial melting occurred—a process that continues today with global warming—the offshore coastal landscapes of the early Holocene period and the archaeological sites upon them were flooded as sea level gradually rose to its current position. As a consequence of this isostatic (land) and eustatic (ocean) history, many of Maine's early Holocene Paleo-Indian and Early Archaic sites are underwater; only shore-side sites younger than 3,500–4,000 years old have been preserved, including many of Maine coastal Indian shell mounds. But today even these sites are suffering from storm damage and erosion as sea rise accelerates.

Quebec lies to the west and north, occupying about 33 percent; and New Brunswick lies to the east of Maine's "hump," occupying 43 percent of its boundary. In other words, roughly three-quarters of Maine is contiguous to a foreign country, and all territory north of 45 degrees latitude is enclosed by Canada.

Maine's northwest border was established as a "height of land," using a chain of mountains to mark the border between Canada and the United States. Subsequently, almost every raindrop or snowflake that falls on the Quebec side of the border enters the St. Lawrence watershed; all precipitation that falls on the Maine side enters the Gulf of Maine watershed. The prevailing southwest winds propelled ships sailing from points south in an easterly direction. Thus, the term "Downeast" came to refer to the region of Maine extending from Acadia National Park through coastal Washington County, or the US-Canada border of New Brunswick.

The St. Croix River, which along with the St. John River constitutes the eastern border of Maine, was the site of the first landing in June 1604 of French settlers intent on establishing a colony for King Henri IV. Pierre Dugua, Sieur de Mons, the Lord of Acadia, and Samuel de Champlain, the King's Geographer, endured a miserable winter on St. Croix Island in 1605 during which 35 of the 79 potential settlers died (Fischer 2008: 173). This short-lived settlement was the first manifestation of New France and, as such, the birthplace of Canada. The French also established Jesuit missions on Penobscot Bay in 1609 and Mount Desert Island in 1613, the same year another Frenchman established the town of Castine. The English were less successful: the branch of the Plymouth Colony settled at Popham, just south of today's coastal city of Bath, in 1607–8 lasted little more than a year; a settlement at York in 1623 also failed.

PREHISTORIC PEOPLES

Maine's Indian history spanning some 12,000 years includes a parade of cultures that began with the Paleo-Indian Clovis culture. These earliest Indians arrived when the Laurentide Ice Sheet was just beginning to retreat from the continental shelf; much of the country was still cloaked in tundra with pockets of spruce and birch in the lowlands and river valleys. Clovis people (ca. 9,500–8,000 BC) were specialized big-game hunters

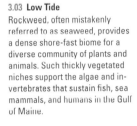

3.03 Low Tide
Rockweed, often mistakenly referred to as seaweed, provides a dense shore-fast biome for a diverse community of plants and animals. Such thickly vegetated niches support the algae and invertebrates that sustain fish, sea mammals, and humans in the Gulf of Maine.

whose spears and darts were armed with fluted points made from chert originating in quarries as far afield as northern New York, Pennsylvania, and even Labrador. They ranged widely across the uplands and the exposed continental shelf, hunting mammoths and caribou in country that resembled northern Labrador today. After the glacial ice retreated, the continental shelves were flooded and the mammoths were hunted to extinction. Eventually, Clovis Paleo-Indians were replaced by cultures with a more diverse economy, known to archaeologists as Northeastern Archaic.

The mixed pine and hardwood forests that had replaced the spruce-tundra parklands of the earlier age supported new economic strategies for Early Archaic cultures (ca. 8,000–3,000 BC). Deer and moose became the predominant prey, augmented by small game, fish, berries, and edible plants and nuts (fig. 3.15). Among the new developments was a mortuary tradition in which leaders were buried with weapons and tools beneath stone mounds seen at the 7,500-year-old L'Anse Amour site in Forteau (fig. 6.10), on the Strait of Belle Isle in Labrador. By Late Archaic times (ca. 3000–1500

< < < CLOCKWISE FROM OPPOSITE BOTTOM LEFT

3.04 Homestead on Jaquish Island
Located off Bailey Island, south of Brunswick, Jaquish is one of many islands along Maine's coast that is large enough for only one home.

3.05 Cobscook State Park
Downeast Maine's Cobscook Bay has a tidal amplitude averaging 24 feet between high and low. The filling and flushing of waters twice a day mixes nutrients from land and sea, creating the rich marine life for which Maine is famous.

3.06 Reid State Park
The southern end of Georgetown Island in midcoast Maine has sandy beaches, ocean surf, and backwater marshes that host a wide array of water birds.

3.07 Maine's Rock-bound Shore
One can imagine these breaking seas on a rugged headland in Reid State Park as a classic Winslow Homer painting.

3.08 Halfway Rock
This barren windswept rock is so named because it is halfway across the mouth of Portland's Casco Bay, making it an important landmark for navigation by both water and air.

3.09 West Quoddy Head
The lighthouse on West Quoddy Head in Lubec is the easternmost point of the continental United States, once the center of a thriving sardine fishery. No longer serving only mariners, the light-house and adjacent state park delight tourists who trek along its clifftop trails.

3.10 Marshall Point Light
The development of electronic navigation aides has eliminated lighthouses in many places in the world. But lighthouses are still iconic symbols of Maine's maritime history and part of its modern identity as a tourist destination.

3.11 Maine's Cultural Landscape
Fishing gear is everywhere apparent along the Maine coast—on a wharf, in a boat, or in the back of a pick-up truck. Each lobsterman's identity and territorial waters are marked by a unique color code on floats and buoys.

3.12 Ruby-Crowned Kinglet
Regulus calendula is only a little larger than a hummingbird, but ranges as far north as the raven. Because of its extremely high heartbeat and inch-think plumage, the kinglet can survive in very cold climates.

3.13 Northern Hawk Owl
The hawk owl (*Surnia ulula*), common in Canada, sometimes migrates south for the winter. This one was spotted hunting mice and other small game in Bristol Mills, Maine.

3.14 Wild Turkey
A few decades ago *Meleagris gallopavo*, the wild turkey, was all but extinct in northern New England, but since reintroduction three decades ago it has become abundant. Often found in flocks of forty or more birds, they are at home in open oak or beech forests and have little fear of humans. Turkeys are rather clumsy flyers and bang about into trees, hence the linguistic phrase, "you turkey!"

BC) the Gulf of Maine ecosystem became highly productive, and its cultures utilized both maritime and land-based resources, developing expertise as deep-sea fishermen and hunters of seals, porpoises, walrus, and swordfish. Mound burials were replaced by large cemeteries at sites including Ellsworth Falls, Orland, Bucksport, Nevin, and Turner Farm, all containing red ocher-covered graves. The dead were equipped with artfully made stone tools and such ritual objects as swordfish-bill daggers and slate bayonets engraved with geometric designs. Some graves contained points made of Ramah chert, a translucent icelike stone originating from quarries in Northern Labrador. During their heyday, these maritime-oriented Red Paint (also known as Moorehead phase) people traveled the coast in large dugout canoes, maintaining contacts with related groups from Maine to Newfoundland and northern Labrador (Bourque 2012). Sharing similar technology, art, and mortuary traditions, and linked by trade in copper, slate, and Ramah chert, the Moorehead-Maritime cultures reached their peak development 3,700 years ago, and then suddenly disappeared.

The final phase of Maine's prehistory began with the appearance of the "broad-spear" Susquehanna tradition about 3,700 years ago. The Susquehanna culture brought new technology in the form of soapstone cooking pots and wide-stemmed points, but their burial traditions and ground stone tools were similar to the preceding Moorehead phase. Whether they represent the arrival of new people or simply the development of a new culture influenced by southern concepts is debated. Their appearance initiated a transition from the previous maritime tradition to a more localized coast- and forest-oriented culture. After a few hundred years, stone pots were replaced by ceramics, dugouts by birch-bark canoes, and marine connections and trade by shell mounds, incipient agriculture, and local adaptations. For the next 3,000 years a sequence of closely related "Woodland" cultures prevailed; in-shore fishing, harvesting of shellfish, and crop agriculture supplemented deer and moose hunting. The arrival of Europeans initially created a lively trade in fur and forest products; but soon European dominance of the coast and expansion into the interior resulted in Indians being displaced from all but a few locations. Maine's remaining tribes, the Mi'kmaq, Abenaki, Maliseet, Penobscot, and Passamaquoddy Indians, sometimes referred to under the collective name Wabanaki (the term of an eighteenth-century confederacy), occupy lands in western and northern Maine. Most have achieved federal recognition and rights to land and resources (fig. 4.02). The Mi'kmaq

3.15 Prehistoric Fish Weir
The remains of this ancient Indian fish trap near Arrowsic Island is visible rarely, only a few times each year at extreme low tides. Fish come in with the tide and are trapped as it ebbs. Such ephemeral features, sometimes seen but mostly invisible, are part of humanity's vanishing cultural landscapes.

Nation, parent of the contemporary Wabanaki, has twenty-nine bands and a population of approximately 30,000, most of whom live in New Brunswick (fig. 4.03). All have struggled for years for rights to land and resources, invoking the federal government's responsibility to implement the Nonintercourse Act, which recognizes Native American claims to land acquired after 1790 by state governments without federal approval. The Passamaquoddy Indian claim was resolved when the Maine Indian Settlement Act (1980) passed the US Congress and resulted in a $81.5 million settlement.

POPULATION

Census records over the years reveal that Maine has enjoyed a modest rate of population increase. But even before Maine attained statehood, some of its people were seduced by the lure of the West to migrate. The first large-scale exodus was probably spurred by the privations of the War of 1812. "Ohio Fever," the expansion of settlement west of the Appalachians into the Ohio Valley, depopulated some fledgling Maine communities and stunted the growth of others between 1815 and 1820.

As the American frontier continued to expand westward, some former residents of Maine who possessed lumbering skills that were particularly prized in the forested states of the frontier, moved on: first to

Michigan, Wisconsin, and Minnesota and later to California and the Pacific Northwest. The California Gold Rush of 1849 provided another opportunity for Maine's woodsmen and a major boost to shipbuilding economies, which supplied many of the ships needed for the West Coast trade: building lumber was shipped around the Horn from Maine until a sawmilling industry was established on the West Coast. Sea-going vessels built in Maine also carried gold-seeking migrants, and thus many Mainers were transplanted to California and the Pacific Northwest.

For fifty years subsequent to the US Civil War (1861–1865), Maine grew slowly, but within one hundred years, its population had doubled. Between 1840 and 1930, roughly 900,000 French Canadians left the Province of Quebec and the Acadian villages of New Brunswick to emigrate to the United States and settle in New England. Many French walked to the United States on the Kennebec-Chaudière Trail, which extended 630 km (230 miles) from Quebec City to Popham Beach (mouth of the Kennebec River). Others migrated by railroad down the same river corridor. Poverty pushed farmers and foresters to Maine, many to the town of Lewiston, which became a center of the Franco-American community. Some immigrants became lumberjacks, but most concentrated in

industrialized areas and into ethnic enclaves known as
Little Canadas, contiguous to an Anglo or Irish community, for example, Waterville and Winslow, Brunswick and Topsham, Saco and Biddeford, Auburn and
Lewiston.

With the French Canadian immigrants came the
Catholic Church, represented by the parish priest;
this religion, along with traditional family life and the
French language, formed the underpinnings of French
culture. The French language was the means for dis-
tinguishing and validating their culture and faith as
separate from that of the Anglican Church established
by the British in North America.

Québécois women began to break away from a
purely housebound life by working in the various
factories and mills, often taking their younger children with them to the factory floor. Franco-American
women were a large and important part of the New
England workforce, willing to accept the cycli-
cal nature of mill work. Although this made their
incomes precarious, women and children helped
Franco-American families survive through their new
roles in labor (Waldron 2005). French American
women often delayed marriage until their twenties,
whereas in Quebec they married in their mid-teens.
When they married, Franco-American women had
smaller families than most women living in Catholic
communities in Canada. Many mill women never
married; they enjoyed life and friendships in the mills
or in the boarding houses, although some eventually
married as they aged. Oral accounts suggest that those
who never married chose self-reliance and economic
independence as reasons for choosing work over
marriage and motherhood.

Most cross-border immigration, until it was virtually eliminated by the Depression, was for economic
reasons. Given the large size of French-Canadian families and a fixed land base predicated on agricultural

3.18 Flower Petal Baskets
Parker is known for crafting flowers from the wood of the brown ash tree, which are woven into her basket designs. She has received the National Endowment for the Arts National Heritage Fellowship, the nation's highest honor in traditional and folk arts.

3.19 Molly Neptune Parker
A fourth-generation basket weaver, Passamaquoddy Indian Molly Neptune Parker has served as president of the Maine Indian Basketmakers Alliance. She is a master teacher in Maine Arts Commission's traditional arts apprenticeship program and demonstrated her craft at the 2006 Smithsonian Folklife Festival.

< < < OPPOSITE
3.17 Logging in Bradford
Some of Maine's forests are so thick you cannot walk through them. Their dense growth has stimulated technology like this feller-buncher, which cuts its way into the dense woods, clearing a path as it goes. No longer an ax and saw affair, Maine woodsmen use computer-assisted saws to prepare wood for market.

use, migration to the mills of New England, aside from working in the woods, was the only viable economic alternative for Canadians. In a 2012 survey, 24.3 percent of Maine's population identified themselves as Franco-Americans, making them the largest single ethnic group in the state (Myall 2012: 3). The Richard family, Will's ancestors, followed this rite of passage to US citizenship, settling first in Vermont and New Hampshire as farmers, loggers, and mill workers. Family and cultural connections continue to remain strong across the Maine–Canada border.

Maine's population also grew with the Rural Renaissance of the 1970s, which was driven by "back-to-the-landers," those individuals who chose not to migrate out of the state for jobs; retirees also returned home to Maine or chose it as a low-cost destination for their golden years. The 1980 census recorded a 13.3 percent increase over the past decade and the growth continued in 1990 with a 9.1 percent increase. Since then, growth has returned to a little below the normal rate of 4 percent established in the 1970s. The US Census identifies Maine as the state with the oldest population and with the lowest population density east of the Mississippi River.

Although payroll and number of employers are still growing in northern Maine, this trend is not great enough to stem outmigration from this region. Has this slower economic activity been a consequence of tighter border controls? Or has the decline of logging activity,

brought on by reduced housing construction and closure of some paper mills, caused this slower rate of growth? From 1996, northern and then central Maine have drawn an entirely new immigrant population, the Amish from the central United States.

ECONOMY

Until 1820, when the District of Maine seceded from the Commonwealth of Massachusetts to become the State of Maine, it was a *de facto* natural resources colony of Massachusetts. Granite was cut from such coastal quarries as Blue Hill, Vinalhaven, Deer Isle, and Stonington (which is still operating today), and shipped to the cities of the eastern US: Boston's Quincy Market, the White House, Washington Monument, and other buildings in the nation's capital; even to Chicago for its Board of Trade building (Allin and Judd 1995). Prior to Freon-cooled refrigerators, ice was cut from the state's rivers and lakes, packed in sawdust, and transported to places as far south as the Caribbean. Blocks of ice were cut from Ice Pond on Monhegan Island for more than a century, through 1974; the tools are preserved at the lighthouse, which is now a museum complex. Ice was often carried as refrigeration to an interim destination where perishables were picked up and then delivered to the cities of the east coast.

The Kennebec and Penobscot river valleys became virtual log conveyor belts from the 1820s–30s until 1976 when log drives became illegal subsequent to a

suit filed by the Environmental Protection Agency. Logging crews penetrated deep into the Maine woods diverting streams and constructing dams to float pine and spruce down to sawmills located at waterfalls. Lumber was then shipped from Maine ports throughout the world. By the early twentieth century, pulp production had spread pervasively into the Maine woods and river valleys, and papermaking became a major industry.

From the earliest days, some of this timber went to Maine's shipyards, where "windjammers," wooden ships powered by wind and sail, were built. Historically, the people of Maine have been hewers of wood, starting with cutting virgin white pine for the masts for King George III's navy. The first ship built in the New World, the *Virginia*, was constructed in 1607 by the Popham Colony. A re-creation of the *Virginia* is being built on the banks of the Kennebec in Bath, Maine, by Rob Stevens who also built the *Snorri*, a modern Viking knarr. Maine's shipyards built an array of ships that through the early twentieth century became increasingly efficient, as measured by speed and carrying capacity. "Down-Easters," ships as long as football fields that were the international container ships of the day, were filled with granite, ice, lumber, or grain for export.

FARMING THE LAND

Aroostook County, the hump that forms Maine's northern boundary, is the most fertile agricultural district of the Maritime Far Northeast. It measures 6,829 mi^2 (17,687 km^2) and is the largest county east of the Mississippi River, about five times the size of Connecticut. It was acquired by the US in 1843 as a consequence of the "Aroostook Potato War" that also established the location of the border between Maine and New Brunswick. Two principles constituted basis for agreement: the British wanted the St. John River as a transportation corridor to Quebec, and the Americans wanted the rich agricultural lands of what became Aroostook County.

Maine remained a largely agricultural state well into the twentieth century, with most of its population living in small, widely separated villages. With short growing seasons, rocky soil, and relative remoteness from markets, agriculture was never as prosperous in Maine as in other states; the populations of most farming communities peaked in the 1850s and declined steadily thereafter. Potatoes remain the agricultural mainstay of the economy on both sides of the border, but while Maine farmed more potatoes than any other state in 1940, when its potato fields were featured on "Vacationland" postcards, it had fallen to eighth place in production by 1994.

Today there is an agricultural renaissance occurring in Aroostook County. While the potato crop has slipped, crop diversification has been rising: vegetables such as broccoli; grains for Maine's burgeoning local bakeries; and hops (*Humulus lupulus*) for the many craft breweries that make Allagash, Geary, Shipyard, Sebago, Gritty McDuffs, Baxter and many other brands important agricultural products. In the early 1990s,

3.22 Amish School
Amish schools teach state-approved curricula, but all schooling is done in a one-room schoolhouse.

3.23 The Pioneer Place, U.S.A.
Located in Smyrna, Maine, in northern Aroostook County, this store owned and operated by the Amish community has a telephone but no electricity. Refrigeration and lighting is by gas; heat is from wood; and the cash register is operated manually. Tools are powered by compressed air, drawn from windmills.

attracted by its cheap and sometimes abandoned farmland, a stream of Amish farmers (sometimes referred to in Maine as Mennonites) began arriving from Tennessee, Kentucky, Pennsylvania, and Indiana. This is a veritable "agri-culture," truly embedded in working the land. Recognizable in their straw hats and women's bonnets and traveling in black horse-drawn carriages, these new arrivals are producing traditional Maine crops as well as entering the carpentry and roofing trades.

Maine agriculture today is a mix of apple orchards, dairy farms, poultry, and specialized products (fig. 3.24). Along the Canadian border maple sap is extracted and transported via plastic tubes, sometimes across the border into Quebec, where it is boiled down to syrup. The economies of ecologies do not recognize arbitrary borders. "Wild" low-bush blueberries are also grown on both the glacial outwash plains of Downeast Maine and adjoining New Brunswick and have been sold through shared marketing endeavors.

Maine is a leader of community markets, places where obtaining one's food is much more than an economic transaction; it is an opportunity to engage in the social and political activities of citizenship. Manifestations of this democratic impulse are seen in the public engagement in green production and distribution systems like Johnny's Seeds (fig. 3.25) and the Maine Organic Growers and Farmers Association product stores.

BOUNTY FROM THE SEA

Alongside its potatoes, Maine is well known for its seafood and fisheries. Primary production of organic compounds by phytoplankton is extraordinarily high in the Gulf of Maine. Its phytoplankton grows "Like grass in a New England hayfield....Biologically the Gulf of Maine is a garden, one of the most productive coastal ecosystems in the world" (Bolster 2012: 22, 66). In a process known as primary productivity, photosynthesis utilizes solar energy and waterborne nutrients to produce more plant matter in the form of zooplankton, small animals like copecods that drift with the current.

Maine's fisheries have, unfortunately, seen better days than in recent decades. Over the centuries, fishermen have contributed to the vortex of fishery depletion by developing more efficient technology to catch the remaining fish faster than the fishery could reproduce itself. Depopulation of the marine life in the northwest Atlantic, the region from the Gulf of Maine to Newfoundland and southern Labrador, began as early as the late sixteenth and seventeenth centuries when Basque and Dutch whalers extirpated many of the large whales from Newfoundland to Greenland waters. While early refinements in fin-fishing techniques such as the jigger, a baitless lure for codfish, had little widespread impact, technological developments of recent centuries have been catastrophic (Bolster 2012: 221–2). In the 1840s, the purse seine was designed to trap herring, but it also caught everything else. In the mid-1850s, the longline with thousands of hooks was introduced. In the early twentieth century the bottom-scraping dragnet was developed. It caught everything in its path, much of which was discarded, and destroyed the seabed that

3.27 Sardine Weir
Sardines, small oily fish in the herring family Clupeidae, are still being caught with weirs constructed of sticks driven into the shallow seabed and connected by nets. Unsustainable harvesting in recent decades caused the sardine industry to crash. In the 1950s, the industry supported fifty sardine canneries; the last cannery closed in 2012.

3.28 Alewives
Alosa pseudoharengus, a small but abundant fish in the herring family, spends the majority of its life at sea but returns to freshwater to spawn. Alewives grow to 10–11 inches in length, weighing about half a pound. While in the Gulf of Maine they serve as a feed stock for larger fish. In May when alewives ascend streams to spawn they become a banquet for ospreys (*Pandion haliaetus*).

in the 1820s, when new mills were established in communities with waterfalls for power and farm girls for labor. Cotton cloth was the primary product of a manufacturing center at Lewiston and Biddeford; Dexter and Vassalboro produced wool. Factories for canning agricultural products and fish—like Snow's Clam Chowder, located near Pine Point (fig. 3.29) because of the easy availability of surf clams, processing of leather and shoes, and paper making all added to Maine's economic base. A dedicated relationship between Maine's forests and its paper mills continued through the 1980s, when the "working forest" became fair game for speculative interests. As lands have been sold for housing and recreational development, the long-time protective relationship between the forest-products industry and forest land has become tenuous. Mill owners have come and gone, and the names of mills and types of paper products have changed; but nevertheless the industry remains strong in Maine because the pay is good. In 2010 the average yearly wage in paper manufacturing was $63,591, which is about double the state's average annual wage (Maine Department of Labor and U.S. Bureau of Labor Statistics, 2010).

SEAFARING CONSTRUCTION

Another backbone of Maine's economy—shipbuilding—used to be linked to its forest resources. Maine's tradition of shipbuilding has weathered four

3.29 Surf Clamming
During very low tides, surf clams are harvested on the clam flats at Pine Point south of Portland. When such tides occur in December it makes cold work for diggers. Snow's Clam Chowder was produced for years from these clams.

3.30 Ancient Call of the Clam
Humans have dug for clams since the dawn of humanity. The earliest-known archaeologist evidence for clamming dates to 125,000 years ago in South Africa. Here a Korean couple who immigrated to Maine carry on the ancient tradition.

hundred years of change and adaptation from wind-driven wooden ships to coal-driven vessels with metal hulls that began to be built late in the nineteenth century. In the forefront of that conversion was a shipyard on the Kennebec River now known as Bath Iron Works, which for decades in both Bath and Portland has been a primary contactor to the US Navy, most recently building ships as guided missile destroyers, cruisers, and frigates. During World War II, Portland was a primary producer of military ships. Since then, it has been used for specialty jobs such as construction of sea-going oil rigs.

Maine's coast, where sea and land interact through the activities of the people, defines Maine's sense of place. After almost five centuries, boat building continues to be the finest expression of Maine craftsmanship—both in wood and steel. Traditional wooden-boat building skills were resurrected by Lance Lee who established the Apprenticeshop Program in 1972 at what has become the Maine Maritime Museum in Bath. Now located in Rockland, the Apprenticeshop offers programs for youth and adults who come from around the world to learn traditional boat-building skills, sailing, and maritime arts. Another equally notable program was born in 1980 in Brooklin, Maine, when local bankers offered Jon Wilson, still in his early thirties, a bargain on the biggest property in the town—a slate-roofed brick "cottage" built for a Boston family in 1916, with a dock and barns on 65 acres at Naskeag

3.36 Windjammer
Maine-built windjammers, designed for long voyages around the Horn in the late 19th and early 20th centuries, were the last wooden cargo ships to be built in the Western Hemisphere. Some carried five or six main masts. They usually carried bulk cargo like lumber, grain, bricks, or ice, and usually followed the prevailing winds while circumnavigating the globe. Several of these ships still sail the coast of Maine as tourist vessels, most centered in Camden.

credited with elevating autumn leaf-turning as the icon of New England (Ryden 2001: 260). Thoreau's genius for word-pictures conveys the brilliance of the season:

> For beautiful variety no crop can be compared with this. Here is not merely the plain yellow of the grains, but nearly all the colors that we know, the brightest blue not excepted: the early blushing Maple, the Poison-Sumach blazing its sins as scarlet, the mulberry Ash, the rich chrome-yellow of the Poplars, the brilliant red Huckleberry, with which the hills' backs are painted, like those of sheep. The frost touches them, and, with the slightest breath of returning day or jarring of earth's axle, see in what showers they come floating down! … These fresh, crisp, and rustling leaves. How beautifully they go to their graves! how gently lay themselves down and turn to mould!—painted of a thousand hues.

Today the observance of fall colors has become a regional industry. "Leaf peepers" cruise New England and the Maritime highways from September through

< < < OPPOSITE

3.38 Breezing Up
Sailing is a sport that draws many to Maine. A spanking breeze brings thousands of sailors out to get a dose of salt spray and sun.

3.37 Kayaking
During the past two decades kayaking has become extremely popular and is well suited to Maine's protected island waterways and runs. This group off Sebasco has little danger of capsizing because fog usually means calm seas.

November to absorb this colorful crescendo before the long winter landscape is unveiled (figs. 3.01, 3.39-41).

Much of the Maine experience is based on outdoor recreation. Maine is not just a coast with beaches, sailboats, lighthouses, windjammers, and kayaks (figs. 3.36–38). The interior is full of mountains, lakes, rivers, and forests where one can experience alpine and Nordic skiing, camping and hiking, canoeing and white-water rafting, hunting and fishing, and just plain enjoying wilderness that has become in short supply elsewhere in the urbanized Northeast. In the 1980s, when Maine started promoting Alpine skiing, markets as close as Massachusetts and New York were surprised to learn that Maine has mountains. New Hampshire considered Maine's gambit to promote its ski resorts (fig. 3.42) as provocative because skiing had long been New Hampshire's recreational niche.

An insightful study of Maine tourism (Lapping et al. 1998) argued that Maine's traditional nature-based industries are not likely to expand, but that tourism would probably grow. From 64,000 recreational visitors to Acadia National Park in its first year of operation, peak visitation in the 1980s exceeded four million, before declining to about 2.4 million visitors in 2011. While

some residents see tourism as a threat to ecosystems as well as to the integrity and health of host communities, when tourist development is properly done, Maine citizens are not disadvantaged by visitors. Tourism supports a broader range of business activities than food and agriculture, forest products, or fisheries because it is more diverse and reaches throughout the state and has linkages to all of the nature-based industries. The authors of this study conclude that policy initiatives should "encourage tourist development patterns that preserve our social and natural heritage while enhancing Maine citizens' economic opportunities and quality of life" (Vail et al. 1998: 76). Fourteen years later, with much of the manufacturing base unlikely to return, the need for a healthy tourism sector is even stronger today.

While Maine still has a healthy pulp and paper production industry, attrition of mills and employment have continued. When mill downsizing began in the Katahdin-Millinocket area in the 1980s, the replacement of well-paid mill jobs with tourism was soundly rejected. A quarter-century later, with a continuing reduction of employment in paper mills, attitudes toward a tourism-based economy in this region have become much more positive. Factors in this transformed

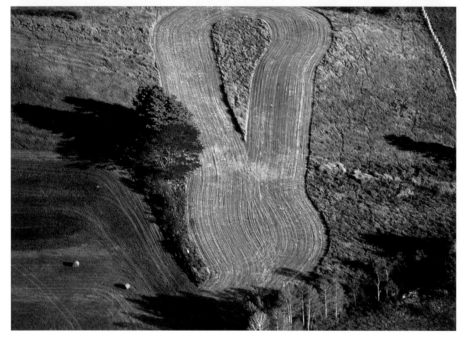

3.39 Wassataquoik Stream Corridor
This path follows one of Maine's waterways, which was formerly used for logging drives. The International Appalachian Trail follows this stream valley east from Mt. Katahdin toward New Brunswick.

3.40 Landscape Features
Two trees, fence, bailed hay, and mowing contours illustrate Maine's agrarian landscape in which farming has been a way of life since the 18th century.

experience, with large retail outlets aggregated in such towns as Freeport, York, and Bangor, all trading on the market image of Maine quality and frugal value. The original and still primary attraction is L.L. Bean in Freeport, Maine, which has become so famous that Canadians and Greenlanders now consider it an essential part of the US experience.

Maine also supports a vibrant cultural life, with art and music festivals, museums and theaters: Portland Museum of Art; the Farnsworth Art Museum (featuring the works of three generations of Wyeths); Monmouth Theater; Ogunquit Theater; the Artist Colony on Monhegan Island; Music Festival and Peary–Macmillan Arctic Museum at Bowdoin College (see pp. 30–31). This state has produced creative talent that has given singular expression to Maine: N. C., Andrew, and Jamie Wyeth, Rockwell Kent, Rachel Carson, Henry David Thoreau, Winslow Homer, Marsden Hartley, Edward Hopper, William and Marguerite Zorach, and Stephen Etnier, as well as contemporary artists Colin Page, Connie Hayes, and Dahlov Ipcar (née Zorach). Maine is also identified with an indelible personality captured in a laconic old-time conversational style noted for brevity and wry Downeast humor. These attributes were captured best by the personalities "Bert & I" as recorded in the late 1950s and 1960s by Marshall Dodge, Bob Bryan, and Tim Sample.

Maine has used its diverse array of natural resources to proactively adapt to the times. Its economy has responded with vigor to changing conditions: from water and wind power to coal power. In 2012, tidal power off Eastport began to be harnessed using a system modeled on a submerged paddlewheel.

In the early 1980s, an ad hoc committee of Maine government and business interests produced an economic portrait of the State of Maine (Colgan 1982). In it, the state was geographically segmented into three economies: south, central, and north. But, somehow through an interpretation not intended by its writers, these three economies became redesignated as the "two Maines," with the south as the location of the "haves" and central and north consolidated as the location of the "have-nots." The study documented an economic configuration separated on a south–north axis along the I-95 corridor. East of the I-95 to the coast were the "have" economies; between the Canadian border and west of I-95 were the have-not economies. Sales tax and employment data support this depiction. These economic patterns are also raising havoc with traditional winter settlements in some areas of the coast where long-time residents are being driven out by rising property taxes and winter residents on islands are being

attitude have included extending the International Appalachian Trail from Baxter State Park to the New Brunswick border, and prospects for a North Woods National Park spurred by large-scale land acquisition by Roxanne Quimby, founder of "Burt's Bees," with funds she received for selling her company. The park would be contiguous to Baxter State Park (200,000 acres), which was created by Percival Baxter through his philanthropic endeavors.

The quality of Maine-made products has also helped spur Maine's economy: clothing, shoes, boots, kayaks, snowshoes and other high-quality outdoor gear that recall memories of Vacationland are particularly popular. Shopping has become part of the vacation

3.41 Salt Marsh and Forest Land
Maine's coastal settlers often had to construct ditches to drain salt marshes before they could grow hay. Traces of ditching can still be seen in the lower part of this image.

3.42 Sugarloaf USA
Maine's prime alpine ski destinations are Sugarloaf, Saddleback, and Sunday River. After years of serving only locals, Maine broke into the Eastern downhill ski market in the 1980s.

3.43 Portland
Looking north from the 86-foot-high Portland Observatory on Munjoy Hill toward downtown. The structure was built in a cow pasture on a ridge in 1807 so ships could be sighted while still at sea, giving Portland's waterfront advance notice of arrivals.

lead)." When the motto was adopted in1820 the intention was expressed in an accompanying resolution:

> as the Polar Star has been considered the mariner's guide and director in conducting the ship over the pathless ocean to the desired haven, and as the center of magnetic attraction; as it has been figuratively used to denote the point, to which all affections turn, and as it is here intended to represent the State, it may be considered the citizens' guide, and the object to which the patriot's best exertions should be directed.

Perhaps a less profound interpretation of *dirigo* is an old Republican Party slogan, "As Maine goes, so goes the nation." Before the New Deal, Maine's presidential primary was held in September, the first in the nation. For many years its outcome seemed to lead the nation by presaging the election outcome, much as Iowa has done in recent years.

Maine is one of the two anchors, along with Greenland at the other end of the arc, that define the land and waterways of the Maritime Far Northeast. Its glacial history and boreal ecology, its landscapes and rocky coasts, its land and sea animals, and its maritime history, industries, and Native peoples link it more strongly with the lands and waters of the northwestern Atlantic. Today as in the past, Maine's future seems to be tied more closely to the developments of its northern neighbors as global warming opens Arctic shipping routes and stimulates industry and population growth.

forced out when their villages lose population (and, hence, schools) after the tourist season.

Another way to measure success is in terms of quality of life, which involves more than one's level of disposable income. The human values of the Maritime Far Northeast accrue from more than economic transactions of the market place. The MFNE is predicated upon the relationship between people and the land. It fosters a spiritual response to the wildness of the sea and forest, winter and summer.

Maine's state seal and flag displays its essential qualities—a pine tree, a moose, a farmer, and a seaman—with the North Star above and the state's motto "*Dirigo* (I

Like Peary and MacMillan, many have found their course northward from the rocky harbors of Maine to those lands and peoples further north. Like other parts of the Maritime Far Northeast, Maine is a small-scale society that depends upon its natural resources and

3.44 Eastern Promenade
The wealth created by Portland's maritime trade enabled sea captains, bankers, and provisioners to build mansions such as these overlooking Casco Bay. Today many of these homes have been subdivided into condominiums.

3.45 Meteor Shower
Viewing a meteor shower on a cold December night from Reid State Park is an experience well worth the discomfort.

3.46 Winter Starkness
Gravestones and a solitary tree on Cemetery Road in North Vassalboro create a winter scene that prompted the following thoughts from the photographer.

A cold clear night without wind is stillness perfected. The sunlight reflected off the moon is further brightened by snow-covered ground. This natural illumination of a deep winter night is like standing under a street light. A white oak in its nakedness becomes a symbol of the deep sleep of winter's hibernation. One can see a universe in the composition of snow, molded by winter's wind.

has survived by grit, good humor, and ability to adapt to change. Like its northern neighbors, Maine's ethic of self-sufficiency, economic restraint, and sensitivity to environment may offer model solutions to problems that much larger segments of the planet now confront.

Having given birth to the Anthropocene, or the Age of Mankind, humanity is experiencing a dramatic change as a result of supplanting the normal geological and astronomical factors as agents of climate change. The modern agroindustrial world is solidly founded on hydrocarbons. The Maritime Far Northeast, that is, the Northwest Atlantic, has two vital roles: First, as the climate continues to warm and glaciers and sea ice melt, new high-latitude sea routes will open between Europe and Asia by way of the MFNE. The earlier whaling, fishing, exploration, and trading histories of this region are likely to experience a renaissance of cross-border activity as a side-effect of the new commercial route. Second, because of the high cost of hydrocarbon-based fuels, the region's environmentally sensitive small-scale cultures from Maine to Greenland are poised to become models for using diverse energy sources with low pollution coefficients. Such a grand design may not work for global civilization, but perhaps it can for smaller regional divisions. Maine—in keeping with its motto, *dirigo*—is heading in this direction.

4.01 Appalachian Mountains, Quebec
The north end of the Appalachian Mountains on the mainland is marked by a lighthouse at the tip of the Gaspé Peninsula in Forillon National Park. The foredge of this mountain ridge is rapidly deteriorating as walls of rock sheer off and drop into the sea.

4 THE CANADIAN MARITIME PROVINCES

The International Appalachian Trail ... establish[es] a long-distance walking trail that extends beyond borders to all geographic regions once connected by the "Appalachian Mountain" range ... to promote natural and cultural heritage, health and fitness, environmental stewardship, fellowship and understanding, cross-border cooperation, and rural economic development through eco-recreation.

—IAT Mission Statement

THE CANADIAN MARITIMES, AS THE lands between Maine and the Strait of Belle Isle (which separates Newfoundland from Labrador and easternmost Quebec) are known, comprise the provinces of New Brunswick, Prince Edward Island (PEI), Nova Scotia, and Newfoundland. These are the smallest territories in the Canadian Confederation, with the lowest provincial populations. Although Quebec is not part of the Maritimes, its Gaspé Peninsula and the eastern part of the Quebec Lower North Shore (see Chapter 5) share much of the same physiography and economy, as well as aboriginal and recent history. Today these lands harbor diverse ethnic populations including several Native American tribes and peoples of English, Scotch, Irish, and French heritage who trace their ancestry here into the early 1600s. A long history of exploitation and settlement by several European nations have made the Canadian Maritimes the most culturally and ethnically diverse region of Canada; for similar reasons it is also the most politically diverse.

In many respects the geology and aboriginal history of the Maritimes parallel that of Maine; many of the geological processes and the cultural histories are similar. Four million years of Pleistocene glacial history created most of the modern landforms, grinding down the highlands and distributing sediments over the lowlands and the seafloor of the Gulf of St. Lawrence. Eighteen thousand years ago during the Late Glacial Maximum, when sea level was one hundred meters lower than today, the land surface of the Canadian Maritimes was three times as large as it is today, and much of the territory was covered by the great Laurentide Ice Sheet.

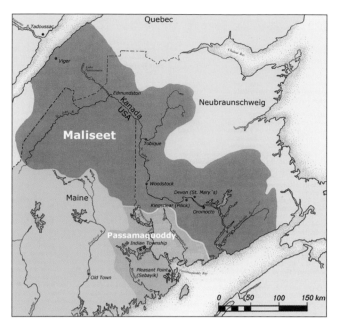

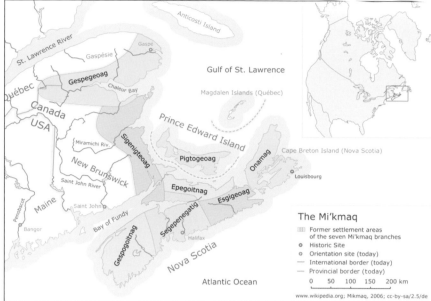

4.02 Maliseet Territory
Maliseet lands included most of the drainage of the St. John River in northern Maine and southern and western New Brunswick.

4.03 Mi'kmaq Indian Territories
The Mi'kmaq are a large and geographically diverse nation that occupied the coastal regions of northern New Brunswick, Prince Edward Island, and nearly all of Nova Scotia during the early contact period.

As glacial ice withdrew, a rapid succession of habitats ensued, beginning with a tundra phase around 12,000 years ago when mammoths and caribou roamed the land and the first Paleo-Indian hunters arrived. In the 1960s, excavations at a large Paleo-Indian site at Debert, Nova Scotia, uncovered a circular arrangement of hearths and fluted point finds suggesting a village encampment. Coastal sites of this age are known from Prince Edward Island but most have been lost to rising sea level. Archaeologists believe that Paleo-Indians were hunting seals and walrus and fishing for salmon at this time, for by 4,000 years ago the early Indian cultures of this region had developed a strong maritime adaptation and participated in the Labrador and Newfoundland Maritime Archaic and the Moorehead Red Paint cultures of Maine and New Brunswick. Ramah chert tools appear during this period and reappear 2,000 years later in Middle Woodland (Ceramic) Period sites here and in Maine.

Further evidence of long-distance trade comes from a 3,000-year-old Adena-style burial found in the Augustine mound near the Mirimachi River in central New Brunswick. The Adena culture is an Ohio Valley culture that preceded the famous Hopewell mound-builder culture. Adena mound burials contain copper and other exotic grave goods. Its mounds, found as outliers as far east as Chesapeake Bay, Long Island, and New Brunswick, are thought to be leaders of a long-distance religious trade cult. The Augustine Mound (Mete-penagiag in Mi'kmaq) burial cache included Adena objects including a copper gorget and copper beads that helped preserve an elaborate headdress and textile garments woven from plant fibers (Turnbull 1976).

Aboriginal connections in the MFNE can be traced by the movement of Ramah chert artifacts and other types of distinctive lithic materials, by the passage of traders across cultural boundaries, and by the migrations of cultures and peoples overland and along the coasts. During precontact times Indian trails thousands of years old crisscrossed the Northeast, and many of these trails still exist. (One that did not exist is the Appalachian Trail, a ridgeline trail established for recreational hikers along the Appalachian ridge from Mount Oglethorpe in Georgia to Mt. Katahdin in Maine. Mt. Katahdin also serves as the southern terminus or beginning of the International Appalachian Trail, which tracks north in Canada.)

Evidence of Native trade argues strongly for a long history of aboriginal maritime-based subsistence and communication. Early European traders took advantage of this Native network to extract furs and other trade products. Later, as European dominance grew and coastal settlements were established at the most productive river-mouth locations, Indian traders were replaced by small-time European maritime traders, and Indians began to be pushed into the interior, losing their maritime subsistence and navigation capabilities gradually as they lost direct contact with the sea and its resources. Eventually they lost their traditional fishing rights and were relegated to portions of the interior that were undesirable for European settlement. Many of these Indian groups never negotiated or signed treaties with the Europeans and found themselves at a loss to protect their aboriginal rights, for which they continue to fight today.

The geography of the Maritimes is dominated in

4.04 Gaspé Storm
A huge storm pounded the south coast of Forillon National Park in 2006 that threatened to wash automobiles into the sea. Gulls profited from the marine life that was hurled onshore by the surf.

the west by the spine of the Appalachian Mountains, which runs through New Brunswick and then disappears, reemerging as the Long Range Mountains of western Newfoundland. These lands are characterized by rolling uplands and glaciated peaks, U-shaped river valleys covered with spruce forests, and tundra and bogs in the higher locations.

Many of New Brunswick's highland regions remain pristine and almost untouched by modern civilization; observed from 40,000 feet they appear indistinguishable from the landscapes of Subarctic Labrador and Quebec. East of these mountain zones are the lowland provinces of Prince Edward Island and Nova Scotia, where most of the agricultural lands, cities, and modern European-derived population are located. During the historic period the European population and economy of these coastal regions—like the Native inhabitants before them—depended on intertidal and deep-sea marine resources. The sandy tidal flats of the Bay of Fundy, Northumberland Strait, and Chaleur Bay produced

immense harvest of shellfish and lobsters; rivers carried large runs of salmon, and the surrounding waters were rich in cod, haddock, flounder, mackerel, and swordfish.

As the earliest-known advance wave of Western European civilization, the Norse reached the New World by island hopping across the North Atlantic. That route, which began in Norway, established a chain of settlements that stretched from the northern British Isles to Faroe Islands, Iceland, and Greenland, eventually reaching Labrador, Newfoundland, and perhaps even points further south on the mainland of North America. The key to Norse success in open-boat ocean sailing was crossing the North Atlantic in the higher latitudes, where the islands and continents are less than 500 miles apart.

One of the first Europeans to visit the coasts of New Brunswick and Nova Scotia was the French explorer Samuel de Champlain (1574–1635). From 1604 to 1607 he explored the coast from Gaspé to central Maine (known to the French as Acadia) and as far

and rocky, derived from glacial outwash and till that has little organic content. The latter make ideal pastures for livestock and for growing bush blueberries. The acid peaty soils, when mixed with sand, are also excellent for growing potatoes, carrots, turnips, and cabbages, which are cultivated in small plots, often along highways. In boggy areas the peaty soil is mounded up in ridges to form elevated beds that facilitate drainage. With sufficient sun, the resulting "dried soil" has enough fertility to support hardy root plants.

All of the Atlantic provinces have until recently been dominated by their forest, maritime, and fishing industries, with strong runs of salmon and other fisheries on its rivers and coasts. For many years Bowaters paper plant at Corner Brook on Newfoundland's west coast, fed by cheap Newfoundland and Labrador spruce, supplied newsprint paper for Halifax, Boston, New York, and Philadelphia papers.

New Brunswick

Maine and New Brunswick are essentially twins. Both geographic entities are about the same size, occupy similar latitudes, and share many of the same rivers, as well as climate, mountain chains, wildlife, forests, and agricultural resources. And its peoples often share the same ancestors. English families in New England who remained loyal to the Crown often escaped the colonies and resettled in the Province of New Brunswick, which was later carved out of Acadia as a home for United Empire Loyalists in 1784 and with Confederation became a Canadian province in 1867.

New Brunswick's major waterway, the St. John River, begins in western Maine and for part of its course forms the international border with Maine, New Brunswick, and Quebec. Then it turns east and south until it reaches the Bay of Fundy through New Brunswick's major city, St. John. The St. Croix, the other river shared with Maine, constitutes much of the border between eastern Maine and western New Brunswick. The province and state also share Passamaquoddy Bay on the western side of Fundy Bay.

Quebec's Gaspésie Peninsula

Two prominent characteristics—a landscape of rock and abundant precipitation—characterize Gaspé. Grossly contorted rock folds known to geologists as synclines (humped) and anticlines (cupped) are dramatically exposed along the North Shore of Gaspé where tectonic plates have repeatedly collided and receded. Cap Gaspé at the tip of the peninsula continues to shear away and drop rock into the sea.

In an average year 57 percent of the days in Gaspé are wet. Given its latitude (48° 58'N) and its average elevation of 1200 m (4,000 ft), the Chic Chocs are a treeless tundra-like plain with a thick ground cover of dwarf herbaceous plants that are hallmarks of tundra vegetation. At these temperate-maritime latitudes, whether in Maine or Quebec, the appearance of Subarctic conditions is largely a function of elevation. As one ascends the hills, the land transforms from northern temperate into Subarctic and, near mountain tops, from Subarctic to tundra. This particular geographic expression of the Far Northeast presents a land dominated by numerous streams and waterfalls, the oceanlike breadth of the Gulf of St. Lawrence, unending hills, tortured rock strata, and dramatic maritime storms. Looming above all are rounded mountains that bear the characteristic signs of having been scoured by the grinding, leveling power of glaciers. Mountain tops were planed off creating flat alpine plateaus. U-shaped valleys carved by glaciers that advanced and retreated during the four million years of the Pleistocene (Ice Age) epoch extend out through the plateau margins onto the coastal plain.

PRINCE EDWARD ISLAND

Located in a southern crescent of the Gulf of St. Lawrence, PEI is isolated from other Maritime Provinces and is Canada's smallest province, smaller even than the State of Delaware. Despite its small size it looms large in Canadian history for hosting the 1867 convention that established the Canadian Confederation. Gastronomically, PEI is known for its red potatoes, Malpeque oysters, and lobster festivals.

Its best-known cultural attraction is the children's novel, *Anne of Green Gables,* written in 1908 by Canadian author Lucy Maud Montgomery based on her childhood experiences in PEI. The novel has sold more than 50 million copies and has been translated into more than 20 languages, making it one of the best-selling novels of all time. The cornerstone of PEI tourism is the green-gabled farmhouse in the town of Cavendish where Montgomery lived as a child.

NOVA SCOTIA

Nova Scotia, which lies exactly halfway between the Equator and the North Pole, was given its name in 1621 when it was awarded to William Alexander by James I of Scotland (aka, James VI of England). This province of proper British manners has a somewhat ungainly physical shape, like a huge, winged bird, and is connected to New Brunswick and the rest of Canada by a broad isthmus. The northern wing is Cape Breton Island, which was connected to the Nova Scotia mainland in 1955 by the Canso Causeway, an engineering marvel constructed in water more than 60 m / 200ft deep.

Nova Scotia's most prominent culture feature is architectural—the huge star-shaped eighteenth-century Louisbourg Fortress, the largest fortified site in North America. Louisburg is located at the eastern tip of Cape Breton Island, the northernmost part of Nova Scotia, and was constructed by the French in 1720–1740 to guard the entry to the Gulf of St. Lawrence and the colony of New France. However, its elaborate defenses were oriented toward the sea and were vulnerable to land attack from the marshes behind. After being captured by a British force in 1745 it was returned to France in a treaty exchange for Madras in India. In 1758 the British captured it again in the Seven Years War, after which it was dismantled.

Throughout the early eighteenth century the French and British fought over Acadia, the French name by which it had been known since Champlain's time. In 1713, after one of these wars in which the English gained ascendency, the French and Métis Acadians who refused to accept English sovereignty were expelled, and many left for other French colonies in New France, stirring Henry Wadsworth Longfellow to write his epic poem "Evangeline" 134 years later. However, by then many of the original French settlers had intermarried with Mi'kmaq and other Indians who were part of the Wabanaki Confederacy and had long supported the French in the wars with Britain. French-Mi'kmaq Métis had ancient roots, intimate knowledge of the land, close Indian relatives, and were Catholic; many of them remained in Nova Scotia. The situation was much more complicated than conveyed in Longfellow's gripping romantic poem of European intrigue.

The ferry crosses from North Sydney, Nova Scotia to Port aux Basques, Newfoundland, a distance of about 160 km (100 mi), but it is a crossing of open ocean, with dangerous currents and frequent storms. Cabot Strait has unpredictable weather, resulting from the mixing of warmer Gulf of St. Lawrence and Atlantic Ocean waters with the cold Labrador Current, and fog is a common companion on land or at sea any month of the year. Sea ice is known to westerners by scores of names—brash ice, slob ice, year ice, multiyear ice, bergy-bits, and many other terms. Inuit people go much further and classify ice in hundreds of varieties. But from Russia and Alaska to Greenland the primary Inuit word for ice is *siku*: "It is our workspace, our picnic area, our classroom, our highway, and our home" (Krupnik et al. 2010: 5).

The basic distinction around Newfoundland, however, is between pack ice and glacier ice, or icebergs. Pack ice is formed at sea from salt water and drifts with

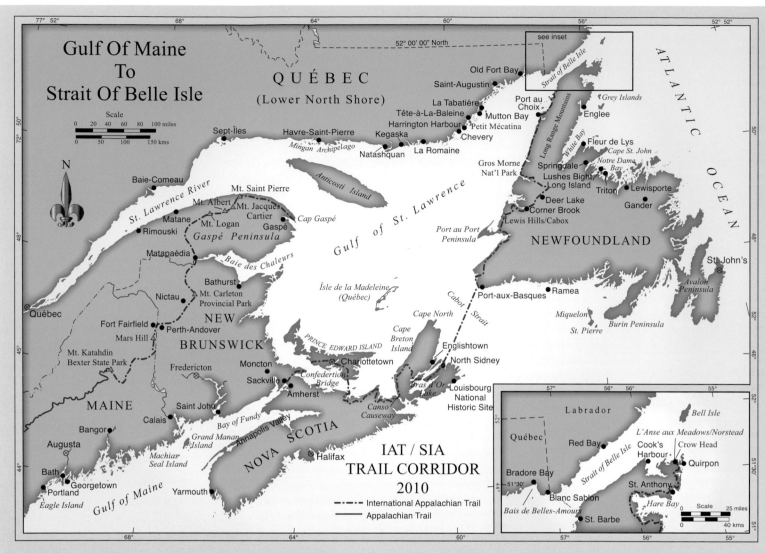

Gulf Of Maine To Strait Of Belle Isle

IAT / SIA TRAIL CORRIDOR 2010

—·—·— International Appalachian Trail
———— Appalachian Trail

4.12 International Appalachian Trail
Beginning at Maine's Mt. Katahdin, the International Appalachian Trail wends its way through the mountains of New Brunswick, Gaspé, Prince Edward Island, Nova Scotia, and Newfoundland.

THE INTERNATIONAL APPALACHIAN TRAIL

By Wilfred E. Richard

THE TERM "BEYOND BORDERS," WHICH connotes common ecological and cultural characteristics on a regional scale, is epitomized by the International Appalachian Trail or IAT (in French, Sentier International des Appalaches). This cross-border endeavor is a tangible symbol of US and Canadian commitment "to work as neighbors to sustain our common environment and to celebrate the grandeur of our common landscape," as its then-president Richard B. Anderson expressed in the 1999 IAT membership application. The landscapes and natural history unite the country through which it passes, and the cultural, social, and linguistic features of its peoples display its diverse heritage. This young international institution already links the lives and lands of the southern portion of the Maritime Far Northeast and continues to extend its reach further north.

In North America, the International Appalachian Trail is anchored in the south by Mt. Katahdin in Maine and in the north by Crow Head in the town of L'Anse aux Meadows, the northernmost point in Newfoundland. Today the trail traverses Maine, the Gaspé Peninsula of Quebec, New Brunswick, Prince Edward Island, Nova Scotia, and the island of Newfoundland. Each regional section of the IAT has its own separate chapter, but chapters cooperate on joint projects and programs and participate in an Annual General Meeting. As a cultural structure based on a natural geographic feature, the IAT fulfills a recreational and spiritual need and recognizes the cross-border nature of the Appalachian Mountains. (A similar cultural and recreational structure is the Maine Island Trail, a coastal small boat trail that links southern Maine to southern New Brunswick.)

The IAT was officially founded when Canadian and American representatives from Maine, Quebec, and New Brunswick met at Mt. Carleton Park, New Brunswick, in July 1995 and were unanimous in their support of a transborder Appalachian footpath. In the fall of 2002, representatives from the Island of Newfoundland proposed extending the trail from the tip of Cap Gaspé, the trail's then-terminus in Quebec, along the western side of Newfoundland to the Strait of Belle Isle. The IAT Board voted unanimously to accept the proposal, and the Province of Newfoundland and Labrador became part of the trail in April 2003.

In 2006, representatives of the Maine and New Brunswick chapters proposed that Nova Scotia be added as a fifth chapter. A sixth chapter soon brought Prince Edward Island into the group. Collectively, as a physical trail, the IAT in North America is 3,016 km (1,862 mi) long and consists of trails, back roads, and abandoned rail bed components, which have been connected by new trails developed by chapter members.

The IAT offers Canadians, Americans, and others the experience of a tangible physical attachment to the land. Originally, the trail was intended to connect the region's highest places: Mt. Katahdin in Maine, Mt. Carleton in New Brunswick, and Mt. Jacques-Cartier in Quebec. But as the IAT evolved that goal and others have been accomplished and more have been imagined. While the IAT is a classic example of a cross-border grass-roots citizen initiative,

the trail is evolving into an institutionalized form of international cooperation. Increasingly, state, provincial, and local governments view the trail as an economic development tool for recreational and heritage tourism. Recognition has resulted in government investments in trail construction and support amenities such as bridges, signage, lodging, sanitary facilities, and marketing. With the decline of the region's traditional ocean fisheries and heavily exploited mines and forests, governments are eager to develop alternative businesses and industries as well as heritage-based travel and tourism.

The geography through which the IAT passes is spectacular. The highest point on the IAT is also the southernmost: Mt. Katahdin, with an elevation of 1580 m (5,268 ft.). Governor Percival Baxter, who purchased the land with his own fortune, presented Katahdin and the surrounding 209,501 acres of land to the people of Maine in 1931. The southern end of the International Appalachian Trail begins on the east side of Baxter State Park and winds 211 km (130 mi) to the north and east, where it enters Canada in the Province of New Brunswick.

Although Maine is larger than the other five New England states combined, compared with its Far Northeast

4.13 Mt. Katahdin
To complete Governor Baxter's plan for a park for the people of Maine, the state in 2006 acquired Katahdin Lake and adjoining lands on the eastern side of Mt. Katahdin. This view of Maine's highest peak is from the summit of Mount Deasey on the International Appalachian Trail (IAT).

4.14 *Inuksuks* on Mount Albert
Built by hikers to resemble stone sculptures made by Inuit in the form of a human (*Inuk*), these cairns were probably erected as art rather than as true navigational aids or spiritual markers:

geographical cohorts, it is small, just three-quarters as large as the island of Newfoundland, for example. Maine's Atlantic coast is 5,600 km (3,500 mi) long and shares the Gulf of Maine with Nova Scotia and New Brunswick. The St. John River is New Brunswick's major waterway. It circumscribes the highlands to the north, then turns east and south, coursing toward the Bay of Fundy.

The New Brunswick section of the trail traverses the northwestern section of that province for 343 km (212 mi) between Maine and the Province of Quebec. Within the IAT trail corridor in New Brunswick and Quebec are many rivers where hikers may change from foot to canoe, assisted by local outfitters who arrange for canoe drop-offs and pick-ups. During summer, throngs of paddlers and fiddlers who make their way downriver in the annual "Fiddles on the Tobique" festival join fishermen along the Tobique. These river valleys serve as part of the IAT route for hiking as well as canoeing. The contiguous region of northwestern New Brunswick and southwestern Gaspé (Quebec) present multiple locations for switching from travel by foot to travel by canoe, where rivers—Restigouche, Tobique, and Matapédia—traverse the same area as trails.

The highest point of the IAT in New Brunswick, and in all the Maritime Provinces, is Mt. Carleton at 820 m (2,693 ft). From here the trail continues northwesterly to the western end of Baie des Chaleurs and enters the Province of Quebec at the confluence of the Restigouche and Matapédia rivers in the town of Matapédia.

QUEBEC'S GASPÉSIE

Upon crossing from New Brunswick to Quebec at the head-of-tide in the Baie de Chaleurs, Québec, the IAT heads into the hills and mountains of the Gaspé Peninsula. Like Nova Scotia, Gaspé juts out into the Gulf of St. Lawrence. With 650 km (401 mi) of trail, Quebec has the second longest section of trail. The trail corridor winds through public parkland, including the Matane Indian Reserve, the Parc de la Gaspésie, the Chic Choc Reserve, passing Mt. Albert, Mt. Logan, and Mt. Jacques Cartier—the latter being the highest point in Gaspé at 1,268 m (4,159 ft). In the middle of the Gaspé peninsula the IAT trail swings north to Mt. St. Pierre on the north shore bordering the St. Lawrence River. Then it dips to the southwest, to a narrow neck of land at Forillon National Park, where what remains of the Appalachian Mountain spine passes into the sea. Some of the peaks of Gaspé, particularly Mt. Albert in the Chic Choc Range, have their summits decorated with *inuksuit* (plural of *inuksuk*), a form of rock art borrowed from Inuit culture of the Canadian Arctic. This area too has an abundance of the necessary building materials—loose rocks of all sizes and shapes.

The trail stops at end of Cap Gaspé, where the hiker returns via the south shore of Gaspé to the small city of Gaspé. There, the hiker picks up public transit (rail or bus) to Sackville, New Brunswick, to access the Confederation Bridge, which connects New Brunswick and Prince Edward Island.

PRINCE EDWARD ISLAND

In 2008 Prince Edward Island (PEI) became the sixth chapter of the IAT. PEI is isolated from other Maritime Provinces and is Canada's smallest province. At its highest point, the island reaches only about 150 m (450 ft.). Nevertheless remnants of the Appalachians exist in its sandy beaches and red, iron-rich soil.

The island's best-known architectural feature, fast becoming PEI's signature landmark, is the 12.9 km (8 mi) long Confederation Bridge that connects it to New Brunswick, and brings hikers to the trail in that province. Walking on the bridge is not permitted; however, there are scheduled public buses.

NOVA SCOTIA

Nova Scotia, which is shaped like a huge, winged bird, is connected to New Brunswick and the rest of Canada by a broad isthmus. From Caribou, Nova Scotia, the hiker passes through farmland, forests, and villages to Cape George and then winds back south and east to Canso to join Cape Breton Island by way of the Canso Causeway, which has connected the island to Nova Scotia proper since 1995. The hiker moves in a northeasterly direction, following the western side of the Bra d'Or Lakes and the St. Andrews Channel, to North Sydney.

NEWFOUNDLAND

At North Sydney, Nova Scotia, the north-bound traveler boards one of Marine Atlantic ferries for a six-hour, 100 km (60 mi) crossing of Cabot Strait, which is known for its unpredictable weather resulting from the mixing of warmer Gulf of St. Lawrence and Atlantic Ocean waters with the cold Labrador Current. Fog is common on land or at sea on both sides of the Strait at any month of the year. The IAT runs 1,200 km (741 mi) from Port aux Basque, one of the old Basque fishing capitals of the Gulf region, located at the southwestern corner of the island portion of the Province of Newfoundland and Labrador, up the west coast of Newfoundland.

Perhaps the most dramatic landscape of the entire IAT is encountered in Bonne Bay—a large fjord midway up the west coast of Newfoundland—and neighboring Gros Morne National Park. Here hikers enjoy spectacular scenery and some of the oldest rocks known on Earth, 1.2 billion-year-old remnants of the earth's inner crust. Visitors can count on close encounters with the largest animal to be found along either the AT or the IAT—bull moose, which can stand seven feet high at the shoulder and weigh more than 1,500 pounds. Moose generally are not aggressive and have calm dispositions; however when cornered or with young they can be dangerous. Enterprising Newfoundlanders have tried to raise young moose whose mothers have been killed in vehicle accidents. These experiments inevitably fail when, as adolescents, their hormones kick in, and the animals become rambunctious and head for the hills.

One has not felt how the wind can blow until hiking the IAT in the hills of northern Newfoundland. Sleet, rain, and snow more often arrive horizontally than vertically. The northern end of the Great Northern Peninsula is rich in history. The small city of St. Anthony, primarily a fishing port, is 40 km (25 mi) north of L'Anse aux Meadows, the much-heralded archaeological site identified as the likely location of Vinland in the Norse sagas. Crow Head, the Newfoundland terminus of the IAT, is visible and a short walk from the Viking site.

Belle Isle is the last Canadian expression of the Appalachians in the Maritime Far Northeast. Administratively part of the Province of Newfoundland and Labrador, Belle Isle lies at the eastern entrance of the Strait of Belle Isle, 37 km (23 mi) north of the village of Quirpon, Newfoundland, and about the same distance from the coast of Labrador. The island acts as an 18 km (11 mi) long plug in the northern end of the Strait of Belle Isle, with tidal currents flowing both ways between the Labrador Sea and the Gulf of St. Lawrence. These waters are turbulent and unpredictable,

for council. This colorful vocabulary, deftly employed by Annie Proulx in *The Shipping News* (1993), has been documented and explicated in the delightful *Dictionary of Newfoundland English* (Story et al. 1990).

Newfoundland is the world's sixteenth largest island, and in 1949 it became the last province to join the Canadian Federation. Because it is such a large island, Newfoundland has many faces. Each of these regions has distinctive landscape or seascape, economy, cultural, and linguistic features. Avalon, a peninsula at the southeastern side, hosts St. John's, the provincial capital, and has a distinct British Isles orientation in its architecture, Scots-Irish music, and linguistic twang. Its harbor has been home-away-from-home to Portuguese and French fishing fleets and during World War II was barred by a huge underwater iron mesh net to make it a safe haven from German U-boats. The highly indented bays and peninsulas of the north coast are home to the island's cod-fishing and sealing fleets. The south coast is a world unto itself, so isolated that even today few of its ports and villages have road access and most access is by ferries and local steamers.

Poor soil and a short growing season limit commercial agriculture in Newfoundland. The Codroy Valley region in southwest Newfoundland is one of the few locations where market-oriented agriculture is possible because an offshoot of the Gulf Stream creates a more temperate, continental climate only in this southwestern corner of Newfoundland. The name Codroy, a contraction of "corduroy"—as in the ridged fabric—comes from a type of road construction used over boggy soils in this part of Newfoundland and is made by laying down a pavement of logs perpendicular to the road's direction and covering it with layers of gravel. Corduroy roads, also still used in backcountry areas of northern New England, have a lifespan of only a few decades before the logs rot and have to be replaced or upgraded.

With the French Islands of St. Pierre and Miquelon only a few leagues offshore, part of its clandestine economy for the past two centuries has been based on smuggling and rum-running. This coast also harbors several local Mi'kmaq communities and reserves. Far from St. John's is the West Coast with its regional center of Corner Brook and nearby the dramatic landscapes of Bonne Bay—a fjord—and neighboring Gros Morne National Park.

Until the 1950s all of Newfoundland outport settlements depended on the fishery, primarily cod, but also lobsters, capelin, mackerel, turbot, as well whales and seals. Fish and lobsters were also major exports, while eastern Newfoundland's produce went

4.23 Robert's Rooms
In Newfoundland, "rooms" are structures or groups of buildings where fishermen store their gear, sleep before heading out to sea and upon returning to port, process their catch for market. These are Robert's Rooms, owned by resident Boyce Roberts.

4.24 Local Cod
Cod—the lifeblood of outport Newfoundland for four centuries—has been scarce around Newfoundland and Labrador for decades, but is slowly making a comeback. For the past few years the government has authorized a small commercial fishery. For Newfoundlanders, life without codfish is marked by economic hardship.

4.25 Fish Flake
The "fish flake," a platform for sun-drying codfish, was the preferred method for preseving cod until salt became available in large quantities in the past century. With the salt-fish method, fresh fish is packed between layers of salt and dried in sheds. Only recently has fresh freezing become possible.

4.26 Trilobite
This magnificent large trilobite, a class of marine arthropods that has been extinct since Paleozoic time, was photographed in Gros Morne National Park. There are more fossils of these species around the world than of any other creature.

east to Europe. Today the fishery is a minor component of outport (that is, small fishing settlement) life except in a few locations like Ramea and François on the south coast and Long Island and Notre Dame Bay on the northeast shore.

Most towns have fallen victim to government centralization policies that discouraged outport life following confederation; but the more immediate cause has been the gradual erosion of the fishery, especially the massive codfish crash that occurred throughout the northwestern Atlantic in the 1980s. In 1993 a codfish moratorium was imposed by the Government of Canada and all fishing—commercial and subsistence—was banned except for a small research fishery needed to monitor changes in the stock. In recent years, as cod stocks began a weak recovery, the ban was relaxed to permit a limited food and commercial fishery.

During this period Newfoundland's largest export shifted from fish to people. With no fishing and little employment opportunity on "the rock," "Newfies"—Newfoundlanders' self-deprecating term for the island's inhabitants—left in droves, seeking work first in Toronto and Hamilton, then in Ontario, and recently in the oil fields and mines of Alberta. By tradition, no one who emigrates ever becomes an ex-Newfoundlander; but even though many return annually or retire to their home villages, for decades the island has experienced a net population loss. During the past few years, following decades of stagnation, Newfoundland's economy has begun to revive as a result of the offshore oil boom. Whether this will revitalize Newfoundland's prospects long-term remains to be seen. A good omen—the first in decades—was a small population increase in 2010.

Although Labrador has been a de facto part of Newfoundland since the late eighteenth century, the association was not formally recognized until December 2001 when the province's name was officially changed to "Newfoundland and Labrador."

The northern end of the Great Northern Peninsula is rich in history. The small city of St. Anthony sits astride a sailing route that was traveled for thousands of years by Native peoples before the appearance of Vikings, Basques, French, and English, and now freighters and supertankers. St. Anthony, primarily a fishing port, has also been important as a staging point for contact with Labrador to the north and as the location of the International Grenfell Mission and its hospital. In recent years its fishing and shipping roles have declined while Viking tourism and activities related to the Grenfell Mission have opened new avenues for the local economy.

For four centuries, the cod fishery served as the primary economic engine of much of the Far Northeast, particularly in Maine, Newfoundland, and Quebec's Lower North Shore, but also in Labrador and Greenland. With the end of World War II, the outport system—of family-based cod fishing with its wind-driven ships, hand jigging from dories, cod traps, and fish dried on flakes—was replaced by much more efficient diesel-powered trawlers and high-tech fishing gear. Modernization of the fleet provided a good income and the fishery remained in the hands of local operators who sold to local middlemen.

Beginning in the 1960s this local fishery began to be overshadowed by an international fleet of factory ships that combined the functions of trawling, processing, and packaging. With this array of technology, in less than a generation the full social and environmental costs were felt as an amalgam of chronic unemployment, abandonment of many outports, outmigration, and closure of the North Atlantic cod fishery. The history of the rise and demise of the North Atlantic cod fishery has been recounted in Mark Kurlansky's book *Cod* (1977) and has been novelized as an economic history in *Sylvanus Now* by Newfoundlander Donna Morrissey (2005). A key issue in Maine and the Maritimes is how to sustain people and resources in an economy that is variously subsistence- or market-oriented. The issue dominates the social and economic transformations taking place throughout the Maritime Far Northeast.

In 2003 the Newfoundland government established a limited "recreational food fishery" program for cod (*Gadus morhua*). For five weeks from August 1, coastal residents who had been forbidden to catch codfish for

4.27 St. Anthony Lighthouse
This lighthouse at the entrance of northern Newfoundland's most secure harbors has been a beacon to sailors for hundreds of years. We caught this view of the light after departing from St. Anthony harbor early one morning in August 2011, in heavy fog.

their own consumption for more than a decade were allowed to take five fish per person per day. Throughout the period of the moratorium the Canadian Department of Fisheries and Oceans allowed an experimental "Sentinal" commercial fishery for small amounts of fish to be taken at designated times and places followed by careful scientific analysis of the catch. By 2012 evidence had begun to indicate that some cod stocks are beginning to rebound, but are still far below their former levels or size.

Quirpon harbor has a deep history, first with northern Newfoundland aboriginal peoples and since the late 1700s as the base for European fishing fleets and a growing local fishing industry. As the *Pitsiulak* casts off from the Quirpon pier, sights are trained on

Belle Isle, 37 km (23 mi) north, at the eastern entrance of the Strait of Belle Isle, and beyond that toward the coast of Labrador and, stretching away to the southwest, the gateway to the Gulf of St. Lawrence and the next stage of our northern journey. The St. Lawrence and the Canadian Maritimes are the geographic core area for the Maritime Far Northeast. Newfoundland and its outport of Quirpon and L'Anse aux Meadows have served as anchors and crossroads for explorations of Leif Eriksson, Jacques Cartier, and John Cabot, and Captain James Cook. Its fishing grounds have provided food and income for generations of European and North American fishermen. Since 2001 this area has also been a focus of the Smithsonian Institution's Gateways research project.

L'Anse aux Meadows Norse Site

IN 1497, FIVE YEARS AFTER COLUMBUS arrived in the Caribbean, John Cabot discovered Newfoundland and Europeans began sailing regularly to Atlantic Canada to fish and hunt whales. The European settlement and exploitation of the Americas had begun—or rather, had begun again.

Cabot would have been surprised to know that Europeans had already explored and settled these same lands 500 years earlier. In AD 1000 Leif Eriksson, son of Erik the Red, sailed south from Greenland, naming the lands he found, from north to south, Helluland (slab rock) for Baffin Island, Markland (forest land) for Labrador, and Vinland (land of grapes, or grass) for Newfoundland and the Gulf of St. Lawrence. Settling at Straumfjord, named because of its strong currents, quite likely the Strait of Belle Isle, he explored, lost his brother Thorvald to an Indian attack, and returned home with furs and grapes. Several years later a larger Vinland expedition was mounted by the Icelandic-Greenland trader, Thorfin Karlsefni, who brought his wife Gudrid. Taking up residence in Leif's "booths" or sod houses, Karlsefni seems to have explored the perimeter of Newfoundland and the Gulf of St. Lawrence and had both friendly and hostile contacts with Native peoples. Leif's and Karlsefni's voyages achieved a landmark of historic proportion. Europeans had settled and explored parts of North America, and although they did not build a colony, Gudrid's and Karlsefni's son, Snorri, became the first European born in the New World. Snorri later became an influential Icelandic leader, and Gudrid made a pilgrimage to Rome where she met the Pope. Back in Iceland, where she worked to establish nunneries, she became known as "Gudrid, the Far-Travelled."

These stories are known from Norse Icelandic sagas that began as eleventh-century oral tradition and were transcribed two hundred years later by Icelandic scribes. Among the many sagas, two—"Saga of the Greenlanders" and "Saga of Erik the Red"—describe Leif's and Karlsefni's voyages. Being unpublished and written in Old Norse, they were unknown outside of Iceland until they were translated and published in the 1830s. The Vinland sagas became an instant sensation in America, and theories about the location of Vinland were soon proposed from Labrador to Chesapeake Bay. But for more than a hundred years there was no physical evidence, and many misinformed and fraudulent claims were made, including a supposed Kensington "rune-stone" in Minnesota (Wallace and Fitzhugh 2000).

In 1961, while searching for Norse sites in Labrador and northern Newfoundland, the Norwegian writer and explorer Helge Ingstad ended a century of speculation. In the small fishing village of L'Anse aux Meadows, Ingstad and his wife, Anne Stine, met a fisherman named George Decker who pointed out some old building foundations in a cow pasture at Black Duck Brook. Upon seeing the ruins, Ingstad immediately recognized the shape of multiroomed Norse longhouses, and excavations uncovered Norse artifacts including a soapstone spindle whorl, a bronze clothing pin, iron nails, and radiocarbon dates of ca. AD 1000. Over the next years Anne Stine's excavations of 1961–1968 revealed a small Norse settlement that included three longhouses, several outbuildings, and a smithy where local bog iron had been smelted to make iron nails. In the smithy they recovered an oval soapstone lamp made by Dorset Paleo-Eskimos around AD 1000, the only item to suggest contact between the Norse and the Native people whom the sagas called "skraelings" (poor or wretched beings).

4.28 L'Anse aux Meadows Following the archaeological excavations in 1961–1968 that revealed the remains of the only known Norse site on the mainland of North America, Parks Canada re-created a sod house village and built a museum to serve as a public interpretation center.

4.29 A Viking Village in America A Parks Canada illustration of the site reconstructs the boatyard, homes, and workshops of this pioneering European settlement in America, built 500 years before Columbus sailed.

5.01 Archaeological Site at Hare Harbor, Petit Mécatina
Less than an hour from Harrington Harbour is the Basque site at Hare Harbor, also known to local French speakers as Eskimo Bay. In 2001, Bill Fitzhugh noticed the rock overhang and judged it to be a likely archaeological site. Ten years of excavation have proven that it was occupied by Maritime Archaic Indians, Groswater and Dorset Paleoeskimos, Basque whalers, Inuit, and French traders.

5 QUEBEC'S LOWER NORTH SHORE— GATEWAY TO NORTH AMERICA

Our harbor is the very representation of the bottom of a large bowl, in the centre of which our vessel is now safely at anchor, surrounded by rocks fully a thousand feet high, and the wildest-looking place I ever was in.

Audubon's 1831 description of Hare Harbor

THE GULF OF ST. LAWRENCE REGION, known to sixteenth-century Basque pioneers as Grand Bay and to modern residents as the Quebec Lower North Shore (LNS, or Basse-Côte-Nord in French), was one of the first regions of North America to be colonized and exploited by Europeans. At first glance this glaciated, hard-rock region would not have appeared promising to the European settler. Exploring the rocky northern shore of the Gulf in 1534 (fig. 5.01), Jacques Cartier recognized that survival here would require persistence, noting it has "hardly a cart-load of earth … I am inclined to regard this land as the one God gave to Cain" (Morison 1971: 354). Indeed, the challenge of making a living from such a hardscrabble land is evident; its coastal towns and settlements cling to rocky hills with little vegetation other than patches of muskeg, grassy tussock bogs, occasional clumps of willows, pockets of spruce trees in the river valleys, and nearly impenetrable tangles of ground-hugging dwarf spruce locals call tuckamore.

The geographic concept of the Lower North Shore is of recent origin. During the nineteenth century the cold-water, barren, rocky coast that stretches from the Moisie River on the LNS to Cape Chidley at the northern tip of Labrador was known to explorers, fishermen, and ornithologists (including John James Audubon) as "The Labrador." The southern part of this region was initially exploited by Basque whalers in the sixteenth century. Following the decline of commercial whaling in the Straits and Gulf after 1600 the coastal waters north of Newfoundland became the province of Dutch whalers while the LNS became a kind of no-man's land occupied by small-scale French, Quebecois, and British fishers and furriers who occasionally collaborated with or fought each other and the Innu and Inuit who lived here (Trudel 1980; Turgeon 1994; Belvin 2006). A dispute between Canada and the British colony of

5.02 Viking Visits the *Pitsiulak*
Bill Fitzhugh, director of the Arctic Studies Center and the Smithsonian's Gateways project, greets Viking reenactor Wayne Hines when the *Pitsiulak* was docked in St Anthony's Harbor, near L'Anse aux Meadows.

5.03 Research Vessel *Pitsiulak*
The Smithsonian's archaeological research from Newfoundland to Baffin Island would not have been possible without the support of a sequence of research vessels. *Pitsiulak* has been our home and research base since 1977. The vessel was originally built by geologist Stearns A. Morse for geological work in northern Labrador. Here she is seen under renovation in Triton, Newfoundland.

Newfoundland over the western boundary of Labrador was resolved in 1927 by the British Privy Council with a decision that created the Lower North Shore (LNS), awarding Quebec a corridor of coastal and interior land along northeastern shore of the Gulf of St. Lawrence. This narrow eastern extension of the Province of Quebec is still sometimes referred to as "Quebec Labrador." The eastern and western borders of Labrador were also modified at that time to their present boundaries.

Despite being geographically close to major population centers in New England and Eastern Canada, the LNS is almost unknown to Americans as well as to many Canadians. Ornithologists know it because of one of its most famous early visitors—John James Audubon—who explored this forbidding coast from Natashquan to Blanc Sablon by schooner in 1833 with his son, capturing and painting birds. En route he stopped briefly at Hare Harbor, the site of the Smithsonian's archaeological research project described below. During this trip he gathered natural history specimens, collected seventy-three bird skins, observed ninety-three different species of birds, and prepared twenty-three bird drawings and paintings, which he sketched and colored "to life" (Townsend 1917: 135). His sojourn included encounters with the Innu, ferocious storms, swarms of

5.04 Harrington Harbour
Harrington Harbour, previously known as Hospital Island because it was the location of a Grenfell Mission Hospital, seen with dramatic light from an approaching storm. With its wooden walkways and colorful homes, Harrington is a picturesque, old-style village.

5.05 Perry Colbourne
A native Newfoundlander, Perry has spent summers for the past two decades as skipper of the *Pitsiulak*.

insects, and other physical and emotional trials noted in his diary. His journal of the summer ends, "Seldom in my life have I left a country with as little regret as I do this" (quoted in Townsend 1917: 146; see also Govier 2003). Audubon's trials had their rewards as judged by the many Labrador bird paintings in his magnum opus, *The Birds of America* (Audubon 1827–1838).

Today's visitors experience many of the same hardships Audubon endured even when fortified by modern maps, travel books, and the Internet. Road travel along most of the Lower North Shore is not possible: there is no highway between Natashquan and Old Fort, a distance of almost 400 km (250 mi.). A ferry serves the roadless coastal villages once a week, but only during ice-free months. In the colder months, ground transportation is largely limited to snowmobile over sea ice or land. Considering its relatively southern latitude (50°N), it is one of the most inaccessible coasts in the Northeast. Other than a few villages hidden in bays and river valleys, this portion of the coast looks as wild and unchanged today as it did to Audubon in 1833 and probably to the Vikings who explored it from their base camp at L'Anse aux Meadows in northern Newfoundland 1,000 years ago.

5.06 Sunrise over Quirpon
Moving through the tickles (narrow channels) at the northern tip of Newfoundland is often a challenge, although the seagulls silhouetted against the rocky horizon backlit by an orange sunrise make the rugged coast look like a pastoral scene.

PREHISTORY

The prehistory of the LNS began around 10,000 years ago as the Ice Age glaciers retreated into the Quebec-Labrador interior and subsequent flooding of the Gulf of St. Lawrence lowlands caused Paleo-Indian hunters to withdraw into the coastal lands that exist today. Confronted by a rapidly changing climate and environment, Paleo-Indians shifted from hunting Pleistocene mammoths and other game to a new coastal way of life: hunting caribou, seals, and walruses with different technology, including boats and harpoons. In the process their culture was transformed into what archaeologists call the Maritime Archaic. Between 8,000 and 4,000 years ago this cultural tradition grew more adapted to marine resources; boats and Eskimo-like toggling harpoons were developed to hunt seals, walrus, and possibly even small whales. By 4,000 years ago they were voyaging from the Gulf to northern Labrador, where they obtained Ramah chert for tool-making, and to Newfoundland, where they traded for slate axes and gouges used for making dugout canoes. Copper, birch bark, and other materials were also traded, and objects made from these exotic materials have been found in Maritime Archaic cemeteries from Labrador to Maine (see Chapters 3, 6). While the disappearance of the Maritime Archaic culture about 3,500 years ago remains unexplained, some combination of cooling climates, southward-moving Eskimos, and the appearance of new groups of Indians from the south contributed to its demise.

The Maritime Archaic tradition was followed by a sequence of Indian cultures ancestral to the Algonquian-speaking Innu who utilized the region's interior and coasts for the next 3,500 years (see Chapter 6).

About 2,500 years ago the first Eskimo peoples, known as Groswater Dorset, migrated from Labrador into the northeastern Gulf and remained for a few hundred years, until they were replaced by Indians similar to the modern Innu, formerly known as Montagnais-Naskapi (Pintal 1998). About 1,000 years ago Vikings based at L'Anse aux Meadows (figs. 4.28–4.31), the Vinland gateway site (Wallace 2000a), explored the Gulf but did not settle and left no saga accounts that can definitely be ascribed to this coast. Five hundred years later, ca. 1530, Basque whalers arrived and for the next hundred and fifty years exploited the Gulf's whales, seals, and fish on a seasonal basis (Tuck and Grenier 1981, 1989; Proulx 1993, 2007; Loewen and Delmas 2011). Basques were followed by other Europeans who began to settle the coast as homesteaders and developed economies based on fishing, trading, and fur trapping. Many of these new arrivals in southern Labrador and the LNS married Native women who helped them learn how to deal with the region's harsh climate and started families based on a blend of European and Innu or Inuit tradition.

Following its early exploration in the sixteenth century, the Lower North Shore slipped into relative obscurity as the wave of European settlement passed through this gateway region to the west, leaving behind small, dispersed communities of Native and European people whose hunting, trapping, and fishing economies remained largely unchanged well into the twentieth century. Until recently historians and archaeologists paid little attention to this region, despite its importance as a geographic transition zone and a migration corridor between Native peoples of the Arctic, Subarctic, and Temperate zones (Frenette 1996). LNS history became even more dynamic after 1500 when its native cultures began to interact with the newly arrived Europeans of various ethnicities and nationalities.

THE GATEWAYS PROJECT

After more than twenty years exploring Labrador's 8,000 years of Indian and Eskimo prehistory (Fitzhugh and Lamb 1985; see Chapter 6) in 2001, encouraged by evidence of Native settlements on the LNS discovered by Rene Levesque, a Jesuit priest turned archaeologist, and of Basque whaling sites that Selma Huxley Barkham had discovered, I expanded the Smithsonian's archaeological research into the northeastern Gulf of St. Lawrence. We intended to investigate the southern geographical limits of the Indian and Eskimo cultures we had uncovered previously in Labrador. Our goals included establishing the southern limits of the Groswater, Dorset, and Inuit (Eskimo) cultures, identifying the factors associated with their expansions and contractions, and exploring their relations with Indian groups and, following 1500, with Europeans.

The Gateways Project would not have been possible without a boat to provide transportation to remote locations while also serving as a warm and secure home base, free of flies and the difficulties of maintaining shore camps. The Smithsonian's research vessel *Pitsiulak* (the Inuktitut word for black guillemot or sea pigeon) was built by geologist Stearns A. Morse for research on the northern Labrador coast. It has figured in our research program since the late 1970s when we were working far from village support

REVEREND ROBERT BRYAN: THE QUEBEC LABRADOR FOUNDATION

5.07 Bob Bryan—humorist, bush pilot, archdeacon, founded the Quebec-Labrador Foundation.

THE VENERABLE ROBERT BRYAN, ARCHDEACON of the Anglican Church, and his wife, Trish Peacock, are ministers who serve their congregations on the Lower North Shore from homes in Harrington Harbour and Bulwer, Quebec. While a student at Yale University, Bob teamed up with folklorist Marshall Dodge to produce the incomparable *Bert and I* phonograph records, which celebrate Maine oral history and its old-fashioned, understated humor. With the proceeds from their initial album, Bob was able to buy his first plane and has since logged more than 17,000 hours of flying time along the North Atlantic coast from Massachusetts to Labrador.

In 1961, Reverend Bryan founded the Québec-Labrador Foundation (QLF), which he continues to serve as Founding Chairman. QLF is a nonprofit organization in the United States and a registered charity in Canada with a mandate to work "beyond borders" (www.qlf. org). Its mission—to foster stewardship of cultural heritage and the environment—epitomizes the concept that inspired this book. QLF focuses part of its efforts on rural communities throughout eastern Canada and New England; it emphasizes local participation and conducts much of its work using interns and volunteers drawn from schools and universities across Canada and the United States. For more than twenty-five years, QLF has also sponsored international exchanges of conservation leaders, initially in the northwest Atlantic Region but increasingly in Europe, Latin America, and the Middle East.

Voyaging to Quebec

By Wilfred E. Richard

5.08 Whale Tail
While crossing the Strait of Belle Isle, the *Pitsiulak* encountered a pod of humpback whales that were churning up the waters in pursuit of krill. Each whale can be identified by the pattern of barnacle encrustations on its tail flukes.

E ACH YEAR SINCE 2001 GATEWAYS RESEARCHERS, including myself, rendezvous by plane or car in Deer Lake, Newfoundland, where we stay at the home of Greg Wood, now a professor at Sir Wilfred Grenfell College, a branch campus of Memorial University in St. John's, who specializes in adventure tourism. The next morning a two-hour drive to the east brings us to Triton on Notre Dame Bay, where the Smithsonian research vessel, Pitsi-ulak, is maintained by our skipper Perry Colbourne, who lives nearby.

After a few days of outfitting the Pitsiulak we set out for Quirpon, a harbor on the tip of Newfoundland's Great Northern Peninsula. As we cruise north up the east side of the peninsula we harbor overnight in the towns of La Scie, Fleur de Lys, or Englee. At the Fleur de Lys soapstone quarry (figs. 5.10, 5.11) I photographed the 1,500-year-old bowl-shaped negative impressions where Dorset Paleo-Eskimos had excavated cooking vessels and lamps. Weather is always our first concern: changes in wind direction, storms, tides, and currents determine our every move. All aboard soon realize that we are no longer on a 9-to-5 schedule; time constraints of the academic and business worlds quickly vaporize.

Heavy seas often batter the 50-foot-long *Pitsiulak* as we travel these coasts, and each member of the crew develops techniques for coping with seasickness. I chew on pilot biscuits (hard tack), a staple of seafarers that still can be found on most grocery shelves in Maritime Canada.

Upon reaching Quirpon, we met Boyce Roberts at the town pier. At Boyce's place (Robert's Rooms fig. 4.23), there followed a sumptuous dinner of fish, moose, potatoes, salads, and beets, accompanied by Newfoundland screech—a local rum with nefarious history—chilled with ice chipped from an ancient iceberg. We visited the L'Anse-aux-Meadows Norse site. Adjacent to L'Anse-aux-Meadows is Norstead Viking, where the Norse knarr, *Snorri*, which was built in Maine, is on permanent display (see chapter 3).

Bill and Perry waited for the right conditions before leaving Quirpon, and dicey weather meant we could go only as far as Cook's Harbor, named for Captain James Cook, who made the first charts and nautical surveys of this region in 1763–68. This is our last harbor before leaving Newfoundland to cross the treacherous Strait of Belle Isle. Along the way, we encounter northern fulmars (*Fulmarus glaciallis*), northern gannets (*Morus bassanus*), Horned puffins (*Fratercula cornicullata*), black guillemots (*Cepphus grille*), and many other birds (figs. 5.12, 5.16–17) as well as killer whales (orcas, *Orcinus orca*), humpback whales (*Megaptera novaeangliae*), and white-beaked dolphins (*Lagenorhynchus albirostris*).

5.09 Brador Fishermen's Harbor
The waters of the Strait of Belle Isle, where nutrient-rich Arctic currents mix with Subarctic waters, is one of the most productive fishing grounds in the MFNE. Here fishermen process mackerel.

5.10 Fleur de Lys, Newfoundland
The village of Fleur de Lys on the western tip of the Baie Verte Peninsula is tranquil in early evening as street lights come on and fog begins to settle in. Soapstone was quarried at this site by the Dorset Paleoeskimos 1,500 years ago.

5.11 Soapstone pot quarry at Fleur de Lys
Round and rectangular blocks were carefully cut and pried from this soapstone quarry in Fleur de Lys, Newfoundland. The blocks were then hollowed out into pots and lamps that were traded to other Dorset groups in Newfoundland for 500 years.

on the northern Labrador coast. In the 1990s we lived aboard the *Pitsiulak* in Frobisher Bay in southeastern Baffin Island, where we were investigating Eskimo-culture connections with Labrador and Inuit contacts with Martin Frobisher's Elizabethan-era voyages of 1576–78. Since 2001 *Pitsiulak* has provided transportation and logistic support for the Gateways Project on the Lower North Shore. Throughout this period *Pitsiulak* has been skippered by Perry Colbourne (fig. 5.05) of Lushes Bight, a small fishing village on Long Island in northeastern Newfoundland. As a young man Perry had fished with his father and brothers in Labrador and knew northern waters like the back of his hand. More than once he has saved us from maritime disaster!

THE QUEBEC LOWER NORTH SHORE

Historically, the Lower North Shore is truly a region without borders, a crossroads and transition zone of cultures in prehistoric as well as in historic times between Québecois and Acadian French, English, Innu (Indian),

5.12 Takeoff

The northern gannett (*Morus bassanus*) hovers just above the water, building up speed before climbing to cruising altitude near the islands of St. Augustine on the Lower North Shore.

5.13 Bird Life Observations

During Will Richard's travels in the Maritime Far Northeast he has observed these migratory species.

Migratory Bird Species of the Maritime Far Northeast		
Arctic Tern *Sterna paradisaea*	double-crested cormorant *Phalacrocorax auritus*	red-breasted merganser *Mergus serrator*
Atlantic puffin *Fratercula cirrhata*	glaucous gull *Larus hyperboreus*	ruddy turnstone *Arenaria interpres*
black-bellied plover *Pluvialis squatarola*	gyrfalcon *Falco rusticolus*	Sabine's gull *Xena sabini*
black guillemot *Cepphus grylle*	harlequin duck *Histrionicus histrionicus*	sanderling *Calidris minutilla*
black-legged kittiwake *Rissa tridactyla*	king eider *Somateria spectabilis*	semipalmated Sandpiper *Calidris minutilla*
Brant goose *Banta barnacle*	Lapland longspur *Calcarius Lapponicus*	short-eared owl *Asio flammeus*
long-tailed duck *Clangula hyemalis*	long-tailed jaeger *Sterocorarius longicaudus*	snow bunting *Plectropher axnivlais*
common loon *Gavia immer*	northern fulmar *Fulmarus glaciallis*	snow goose *Chen caerulescens*
common murre *Uria aalge*	peregrine falcon *Falco peregrinus*	thick-billed murre *Uria lomvia*
common raven *Corvus corax*	purple sandpiper *Calidris maritima*	white-crowned sparrow *Zonotrichia leucophyrs*
common redpoll *Carduelis flammea*	razorbill *Alca torda*	

5.14 Long-distance Migrants

Black-bellied plover, at a tidal pool surrounded by tall-grass salt marshes. Many birds migrate in summer from Temperate, and even Tropical, regions to raise their young in nutrient-rich northern lands. This unusual group of migrants includes two whimbrels (*Numenius phaeopus*), three black-bellied plovers (*Pluvialis squatarola*) and a short-billed dowitcher (*Limhodromus griseus*).

110

5.15 Puffin Colony on Mingan Island
The noise and wake of the *Pitsiulak* alarms some Atlantic puffins (*Fratercula arctica*).

5.16 Peregrine Falcon
While excavating at Hare Harbor the cliffs above us had a nesting pair of Peregrine falcons (*Falco peregrinus*) that vigorously defended their young by swooping upon us while the young were learning how to fly.

5.17 Arctic Tern
Many Arctic terns (*Sterna paradisaea*) were diving into the water next to the fish plant in Harrington Harbour for food, providing an opportunity to capture a good photo of this bird.

5.18 René Levesque
Québec archaeologist and former Jesuit priest René Levesque also conveyed his vast knowledge of the Lower North Shore—particularly the Mingan Islands—to the Gateways team in 2001.

5.19 Selma Huxley Barkham
This well-known historian and authority on sixteenth-century Basque visits to North America received the Order of Canada Award for her research on Basque whaling in Labrador and the Lower North Shore. In 2001, Selma shared her knowledge of Basque history with us on our initial exploratory expedition to the Quebec Lower North shore.

and Inuit (Eskimo). In *The Forgotten Labrador*, one of the few recent publications documenting this region, Cleophas Belvin (2006: 68) describes how French- and English-speaking immigrants arrived on the LNS from Europe, the Canadian Atlantic Provinces, and the New England states. Today some students from the LNS make return migrations for education at English or French schools in Quebec or at New England schools like Hebron Academy. In the past, the Quebec-Labrador Foundation, through the good offices of its founder, Reverend Bob Bryan, sponsored many of these opportunities.

In 2001, the first year of the Gateways Project, we surveyed the entire LNS from the Mingan Islands and Havre-St.-Pierre in the west to Blanc Sablon at the Quebec-Labrador border in the east. In subsequent field seasons the project concentrated around Harrington Harbor (N50⁰ 30'; W590 26'), and, more specifically, on Petit Mécatina Island (fig. 5.01). Most of the better-protected harbors and river mouths have towns or villages. Generally, each of these communities has a distinctive ethnic composition, French, Innu, Inuit, or Newfoundland-English, or some mixture of these. The Innu—who are mostly Roman Catholic—are the majority in the village of La Romaine; Harrington Harbor is largely Newfoundland English; St. Augustine is a mixture of Innu, Inuit, French, and Newfoundlander; and Blanc Sablon is largely Québecois. Havre St. Pierre is one of the LNS villages with a population whose ancestors were among the Acadians forcibly displaced in 1755 (see Chapter 4); they still fly the French flag and retain distinctive musical and literary traditions.

The Lower North Shore is a maze of islands. Unlike Labrador, where the islanded coast provides protection for coastal navigation through a deep-water channel between the islands and mainland, the LNS island passages are shallow and filled with rocks and shoals. Only experienced skippers with local knowledge dare bring vessels like *Pitsiulak* into these channels. Even today many of these runs have never been sounded or charted in detail. Wrecks attest to the navigational hazards of this challenging coast, which is made all the more difficult by the strong southerly winds that blow onshore most of the summer. For these reasons, ever since the Inuit arrived in the 1600s on the LNS with dog sledges, winter has been the season for visiting and formerly for trade and long-distance travel. Today snow machines have mostly replaced dogs. With good weather and ice, trips of several hundred kilometers (200–300 miles) are considered routine.

BASQUE WHALING

In the 1970s Selma Huxley Barkham, working on obscure sixteenth-century archival records in the Basque villages along the northern coast of Spain, discovered documents describing Spanish–Basque fishing and whaling expeditions to "Grand Bay" in the northwestern Atlantic (Barkham 1980, 1984, 1989). These records included legal contracts, financial records, and court cases involving claims by women who had lost husbands or relatives during the voyages. Descriptions of landmarks and places revealed the locations of activity, chief among them being the town in southern Labrador known as Red Bay. Barkham contacted Walter Kenyon, an historical archaeologist at the Royal Ontario Museum, who visited Red Bay with Barkham and confirmed the presence of Basque roof tiles and remains of blubber furnaces. Among the records Barkham found a report of the 1585 sinking of the *San Juan de Pasajes*, a ship laden with a cargo of 1,000 barrels of whale oil that was cast ashore in a storm. Beginning in the late 1970s James Tuck of Memorial University and Robert Grenier of Parks Canada's underwater team began a decade-long research program excavating the Basque shore sites and the wreck of the *San Juan* (Tuck and Grenier 1989; Grenier et al. 2007). Barkham's historical studies and Tuck's and Grenier's archaeological work revealed a previously unknown chapter in the history of northern North America.

In their homelands in northern Spain and southwestern France, Basques today live as sheepherders, restaurateurs, bankers, and retired revolutionaries. (After fighting ceaselessly for several decades for autonomy from Spain, most Basques continue to press for freedom, but do so largely through the political process.) During the medieval period they were expert shipbuilders, ironworkers, and fishermen who pioneered hunting of the large whales that were abundant in the Bay of Biscay during the fourteenth and fifteenth centuries. Following the English discovery of untapped marine resources in the western Atlantic, Spanish Basques established a large shore-based whaling industry in southern Labrador and the Gulf of St. Lawrence that flourished from 1530 to 1600. In addition to Red Bay, which was the largest of the Basque whaling stations, archaeological evidence of Basque activities has been found throughout the Gulf of St. Lawrence from Iles aux Basques near Quebec City to Chateau in southern Labrador. Basque stations were also established on Newfoundland's west coast at Port aux Basques, Codroy Island, Flat Island, Red Island, Governor's Island, Benie Island, St. John Island, and Ferolle Island (Barkham 1989; Loewen and Delmas 2011).

Basque whaling succeeded more from an abundance of whales than from the use of sophisticated whaling technology, which was much inferior to that being utilized by Thule Eskimos and their Labrador Inuit descendants in the thirteenth to eighteenth centuries. In the Bay of Biscay Basques hunted the Greenland right whale (*Eubalaena glacialis*), which probably contributed to the decline of this species. Right whale population today is thought to be less than one hundred individuals but recently has been seeing small gains. In Labrador and on the Quebec LNS the bowhead whale (*Balaena mysticetus*) rather than the right whale was their principal quarry (McLeod et al. 2008). In the late 1500s, after Labrador and Gulf whale stocks declined from overhunting, Basques turned to fishing for cod, which was abundant, easily preserved, and highly marketable in Catholic regions of Europe. During the Middle Ages the Catholic Church expanded fast days so that nearly half the year were "meatless," or days of abstinence, and this created a huge market for Basque codfish (*Gadus morhua*) (Fagan 2006: 147; Kurlansky 1997).

In more recent times, the economic connection between the LNS and the outside world has been through its finfish and harp seal (*Phoca groenlandica*) fisheries. After a moratorium on cod and seals in the closing decades of the twentieth century, the primary fishery resource became shellfish, primarily lobsters, scallops, snow crab, and shrimp. In 2002 Canada's Department of Fisheries and Oceans (DFO) reopened the harp seal harvest with quotas. It was argued that the growing population of harp seals was consuming

cod stocks. With the fishing industry decline throughout the Maritime Far Northeast has come increased awareness of the need to develop heritage- and recreation-based tourism. The Smithsonian's Gateways Project contributes to this aspect of the region's modern economy by providing documentation of a little-known history.

ARCHAEOLOGICAL SURVEYS

As the *Pitsiulak* cruises the coast, we stop at small ports and islands that look promising in terms of ancient site locations or cultural landscapes. Passing shores with good harbors and hunting and fishing locations, we anchor and go ashore for surveys and test excavations. We strike up conversations with local people who know their home regions intimately and frequently provide us with information about old settlements or artifacts. In 2001 one of these conversations led us to Huey Stubbert and the large cache of Ramah chert artifacts he found while digging a garden at his home in Kégaska. Ramah chert is a very distinctive rock that outcrops in Ramah Bay, northern Labrador, and could only have arrived on the LNS with people who had visited the quarries themselves or obtained the stone by trade from Labrador intermediaries. For thousands of years, aboriginal peoples throughout much of the Far Northeast utilized Ramah chert for tools and hunting weapons (Loring 2002). Evan Hadingham (2004: 98) has called this fine-grained flint and chert "the Stone Age equivalent of the Swiss Army Knife."

Toward the end of our 2001 survey we anchored in Hare Harbor on Petit Mécatina Island, located approximately 25 km (15 miles) east of Harrington Harbor and 240 kilometers (150 miles) from the Strait of Belle Isle. The site is marked on the north side by a conspicuous cliff whose overhanging base provides protection during inclement weather. As soon as we landed we found the same type of clay roof tiles used by Basques at whaling sites like Red Bay, just across the Quebec border in Labrador. Basques used these tiles to construct roofs over their workshops and tryworks where whale blubber was rendered. The tiles were spilling from the eroding bank along the shore, and a test pit dug in the clearing below the cliff revealed more tiles. Clearly, Basques had been present at Hare Harbor, but what their operation was like would take nearly a decade of excavations to ascertain.

THE HARE HARBOR SITE

The Gateways Project is based on the premise that this corner of the Maritime Far Northeast has been a significant, though largely unacknowledged, arena in the early history and settlement of Northeastern America. Our exploration of Hare Harbor on Petit Mécatina Island presents a microcosm of this historical panorama and documents some of the cycles of culture contact that have occurred in this region. When we discovered the site in 2001, we did not realize we had been paralleling, 167 years later, the track of Audubon's 1833 voyage from Natashquan to Blanc Sablon. Arriving, we found many of the same birds he

5.25 Harrington Harbour Friends
Among the people who befriended the Gateways crew in Harrington Harbour include (left to right): Claude Ransom, Paul Rowsell, Allen Rowsell, and Larry Ransom.

observed, including another pair of peregrines being hassled by a brood of ravens. What Audubon did not realize—and we soon discovered—was that the grassy meadow below the peregrine's cliff-side nest held 4,000 years of prehistory.

Ten years after Audubon's visit Samuel Robertson published a paper based on a lecture he gave to the Literary and Historical Society of Quebec in 1841 (Robertson 1843: 28). Part of the paper provides information on a site known to local people in the village of Tête à Baleine (Whale Head) as Eskimo Bay, the same place English-speaking people refer to as Hare Harbour.

> there is no want of remains of buildings and tumuli [stone Inuit graves] of such ancient date [pre Columbian], that tradition ascribes them to the Esquimeaux, which in one instance, at least, was false: this occurred three years ago, where a person had occasion to remove part of a 'Terasse,' to make a garden. He found an iron instrument, of about eighteen inches in length, of a crooked form, which I conjectured to be a Cerp, such as were used [as vine-pruning tool?] 300 years ago in Spain—if my supposition is right, the remains must have been those of the Basques, as the Norman and Breton countries are not vine countries.

Although Robertson's "cerp" was more likely a whale blubber butchering tool, or one that had been used by Inuit residents of the Hare Harbor site (see below), Robertson was the first to recognize the importance of Basque as well as Inuit history on the Lower North Shore.

BASQUES AND INUIT

Our excavations at Hare Harbor uncovered a stone-floored cookhouse cluttered with roof tiles and with a large hearth pit at one end. For the Basque journey west to the New World, roof tiles served two functions, first as ballast for their ships and then as roofing material over their workplaces ashore. We also found glass trade beads, clay pipes, and ceramic vessels that dated this floor between 1675 and 1725 (Herzog and Moreau 2004, 2006), a time frame that most archaeologists consider too late for Basques in the northwest Atlantic because most Basque sites in Canada dated to the mid/late sixteenth century. In other areas of the site we found more tiles, hand-wrought iron nails, whale baleen, glass shards, musket flints, musket balls, part of an Inuit cooking pot, decorative glass beads, and a cup-shaped Basque oil lamp. A boggy area nearby contained large numbers of wooden barrel staves, a hammer, and a broken anchor prong. As we extended the excavation we encountered thick lenses of charcoal and burned wood and the remains of a blacksmith shop. Then, excavating below the stone floor, we found the remains of an Inuit winter house that had been burned and covered by the smithy. Hare Harbor was beginning to reveal a truly complicated history.

Two of the excavated European structures—the cookhouse and blacksmith shop—each date to around AD 1700. On the floor of the cookhouse we found large amounts of Iberian earthenware and north European stoneware mixed together with Inuit soapstone lamps and pots. This alerted us to the fact that Hare Harbor had also been occupied by several Inuit families. These

5.30 Inuit Site and Hare Harbour
Bill surveys two Inuit winter houses the team unearthed near Hare Harbor's Basque structures. The *Pitsiulak* is just visible at the end of the bay.

5.31 Whalebone
While excavating the Basque underwater site the divers found only a few flipper bones, causing the team to wonder, "Where are all the whale remains?" Later, when snorkeling near the head of Hare Harbor, we found scores of whalebones partly buried in the sand. To avoid fouling their anchorage, the Basques had hauled the bones to the head of the bay.

has experienced, when winter sea ice and harp seals and perhaps walrus were plentiful. The Groswater occupation was followed by a short occupation by a Dorset Paleo-Eskimo group about 1,600 years ago. After that, Hare Harbor seems to have been abandoned until Basque whalers appeared during the latter part of the sixteenth century. European cod fishermen, perhaps of Basque origin, arrived around 1700, and it is during this period that three Inuit houses were built on the site, apparently as part of a European (possibly Spanish Basque) cod-fishing station. Evidence of burning suggests that some of these European and Inuit structures were destroyed by fire, perhaps in attacks by other Europeans or Indian groups. This period was one of tumult among the many groups competing for the fish, fur, and marine mammal resources of the northeastern Gulf

Ramah chert, shows that these early Indian cultures penetrated the northern Gulf and maintained close contacts with relatives in Labrador (fig. 5.34). More than a thousand years later, ca. 2,200 years ago, a small group of Groswater Paleo-Eskimo people camped on the Hare Harbor terrace and left small concentrations of artifacts manufactured from chert originating in southwestern Newfoundland. These were the first Inuit people to arrive in the Gulf from the Arctic. Their appearance occurred during the Sub-boreal period, one of several cool climatic periods this coast

MARINE ARCHAEOLOGY AT HARE HARBOR

In 2003, underwater explorations became part of the Petit Mécatina field agenda. Wilson Evans, the wildlife officer for Harrington Harbor, had offered to check out the harbor area adjoining the site and on his first dive found an anchor, Basque pottery, tiles, jugs, bottles, and whalebones, similar to materials we were finding onshore. However, because of the frigid seawater, the underwater artifacts were better preserved. The most striking features of the underwater site are a series

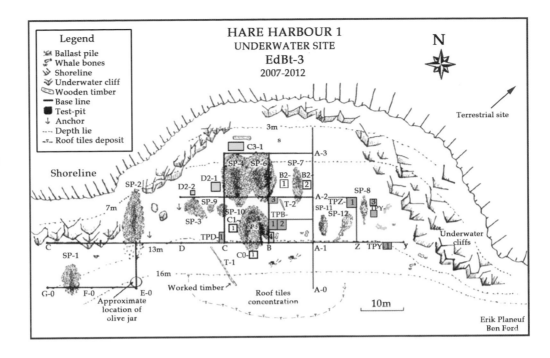

5.32 Marine Dig Site at Hare Harbor
Diver Erik Phaneuf drew this schematic of the underwater dig site showing the location of the Basque ship's ballast piles, which were tossed overboard after crossing the Atlantic from Iberia to North America. Also mapped are locations of whalebones, an anchor, and test pits. Working this site required excavating in bone-chilling Labrador Current water up to 20 m deep.

Legend
- Ballast pile
- Whale bones
- Shoreline
- Underwater cliff
- Wooden timber
- Base line
- Test-pit
- Anchor
- Depth lie
- Roof tiles deposit

HARE HARBOUR 1
UNDERWATER SITE
EdBt-3
2007-2012

N

Terrestrial site

Shoreline

Underwater cliffs

10m

Approximate location of olive jar

Worked timber

Roof tiles concentration

Erik Planeuf
Ben Ford

5.33 Belamine Potsherd
Excavations in 2011 at the north end of the Hare Harbor site proved to be disappointing in terms of artifacts. But at the south end of a contiguous array of 40 two-meter squares, Bill dug a test pit into what proved to be a rubbish midden. this glazed sherd of German Belamine ware dates to circa 1640.

5.34 Artifact Potpourri
A variety of prehistoric stone tools predating the Basque and Inuit occupations was found in 2011 in the Hare Harbor land excavations.

of large, linear-shaped piles of ballast rocks, several large timbers, a concentration of whalebones, and the anchor, which turned out to date to the nineteenth–twentieth century. There was some speculation about whether the ballast piles, which are oriented perpendicular to shore, had been foundations for docks, but subsequent investigation discounted this possibility because no piling timbers were found and the water was too deep for constructing piers and wharfs (fig. 5.32).

Using powerful pumps and dredges allowed us to recover the most delicate finds, even in deep water. Our excavations revealed successive layers beginning with woodworking, followed by butchered bowhead whalebones, and then detritus from commercial fish processing. A cluster of flipper bones showed where whales had been drawn alongside a cliff for flensing and blubber removal. Because these bones were from flippers only, we wondered what happened to the rest of the skeletons. Later, while diving at the shallow end of the harbor, we found the missing ribs and vertebrae (fig. 5.31). The same disposal pattern was found at the Basque site of Red Bay, Labrador, probably for the same reason—to keep the large bones of the carcass from fouling the anchorage.

Hare Harbor is an unusual site because of its early connection with John James Audubon and its long and complicated archaeological history. Additional information about the Hare Harbor excavations can be found in publications by David Malakoff (2007),

119

William W. Fitzhugh (2006, 2009), Fitzhugh and Phaneuf 2012; Fitzhugh (2008, 2011), and Fitzhugh et al. (2011, 2012).

INUIT AND EUROPEANS ON QUEBEC LOWER NORTH SHORE

During the sixteenth and seventeenth centuries contacts between Inuit and Europeans in the Labrador region were often hostile. Excavations at Hare Harbor and other Inuit sites we have investigated along the LNS suggest, however, that by 1700 Inuit and Europeans were finding ways to work together, just as the Basques and Miq'maw Indians had begun to cooperate in southern Newfoundland and northern Nova Scotia. Evidence of European-Inuit interaction at Hare Harbor includes Inuit soapstone lamps and pots on a Basque cookhouse floor found together with iron, tiles, iron nails and other European materials. Because these structures contain Inuit artifacts, and the Inuit houses nearby contain European materials similar to our finds from the cookhouse and smithy, both groups may have been living here at the same time, possibly as partners in a joint fishing operation. The Inuit may have assisted the Europeans in return for receiving valuable materials such as ceramics, iron tools, weapons, and cloth. During the winter when the Basques returned to Europe, the Inuit may also have served as site guards to protect the premises from attacks by rival Europeans or Indians. More work is needed, but our results to date have identified the southernmost Inuit occupations in the world and suggest that a new phase in Inuit-European interaction had begun. According to reports from Martel de Brouague, Courtmanche's successor at the Brador fort, this mutual arrangement did not last very long. He recounted stories of Innu and French attacks on

5.35 Basque Whaling Site
Diver Frédéric Simard brought up bones from bowhead and humpback whales from the ocean floor at the Hare Harbor Basque site.

5.36 Earthenware Jug and Porringer
Frédéric Simard (left) and Erik Phaneuf hold a jug and plate that originated in southwest France. The earthenware jug is decorated in an Iberian style with three strips of squared punch holes that produce a "corn" design. The small porringer made of faience, or earthenware, is glazed with a white tin-based decoration.

5.39 Smithsonian Gateways Team, 2011
The nine Canadian, American, and Mexican members of the 2011 Gateways team pose for a group picture. Lauren Marr, our youngest team member and so presumably fleetest of foot, volunteered to trigger the timer on the camera and then race through three grid squares to the group—within ten seconds. As she dove into place, Vincent Delmas attempted to catch her.

5.40 Gigantic Lobsters
Seafood enthusiasts like to say big old lobsters are too tough to eat, but these granddads were scrumptious, once we figured out how to get them into the pot.

< < < OPPOSITE LEFT
5.37 Chafing Bowl
This unusual glazed earthenware vessel was a surprise find: a rare type of ceramic used to warm pots of soup perched on thee lugs on the rim of a bowl filled with burning charcoal or oil.

< < < OPPOSITE RIGHT
5.38 Ancient Marine Shells
While cleaning out a pond in Harrington Harbor a bed of scallops and mussels was found embeded in marine clay. Radiocarbon dating showed the shells to be 8,000 years old, a relic of the early post-glacial era when the today's land was below sea level.

Inuit families living along the Lower North Shore at Mecatina, Ha! Ha! Bay, and other locations in what seems like a concerted attempt to force Inuit out of the Gulf and back north into Labrador.

RETROSPECTIVE

Ten seasons of sea travel and field archaeology produced many memorable moments for our excavation teams. Discovering old– even ancient—archaeological sites is as fascinating to the student volunteer as to seasoned archaeologists. Archaeology stokes one's curiosity. All that digging is far from boring! Learning how other cultures lived in other times can enlighten and engage the mind. Such questions as, "Who lost that tiny trade bead? Why does this Inuit lamp have a hole punched in it? What kind of games did Inuit children play with toy bows and oil lamps?" inspire the imagination as well as scholarly investigation.

As with any field project, Gateways has had its share of discomforts and stresses, but flying above the site is a gyrfalcon, behind us is the sea, and our skipper Perry Colbourne awaits our return to *Pitsuilak*, our floating field camp in Hare Harbor. In the end, we'd have it no other way.

6.01 Nain, Labrador

The settlement of Nain, the northernmost village in Labrador, originated as the first successful Moravian mission, established here in 1771. Today its population of 1,000 includes Inuit resettled here from former Moravian mission stations further north at Okak, Hebron, and Ramah Bay. The opening of a Labradorite quarry and a mine in nearby Voisey Bay have bolstered its traditional economy based on hunting, fishing, and trapping.

6 LABRADOR – SETTLERS, MORAVIANS AND INUIT

Humans do, habitually and nearly universally, experience a 'something, when in the forceful presence of nature…'. I don't know what that something is. But … I'll call it the wild. The wild is what we don't control. What we cannot fully understand at all past infinitesimal glimmerings… And the crazy, amazing remarkable fact is that it doesn't just kill us with our huge unknowing arrogance. No: in fact, wildness nurtures us, the wild is generative, fertile, fecund, bottomless.

David Oates, *Paradise Wild*

LABRADOR IS A LAND OF GREAT beauty and majesty. It is also the geographic connector between the temperate and the Arctic zones of northeastern North America. Its southern border reaches the Gulf of St. Lawrence, North America's gateway to the continental interior (fig. 6.02). Here deciduous trees and pines are replaced by the coniferous boreal forest that stretches across Canada to Alaska. North of the Strait of Belle Isle, a strip of Arctic tundra created by the chilling effect of the Labrador Current defines the coast. And north of Nain, Labrador's low, islanded shores gives way to the Torngat Mountain chain and subarctic conditions disappear. Trees cede first to shrubs and then to tundra, and by the time one reaches Killinek at the northern tip of Labrador, true Arctic conditions prevail and continue across Hudson Strait, the thoroughfare connecting Hudson Bay with the Atlantic Ocean.

Labrador's great expanse—293,000 km² (113,000 square mi²)—and its nearly 1,300 km (800 mi) north-trending, ocean-moderated coastline ensures that it is also a land of contrasts. Its coast is barren, cool, and often foggy in summer, and in winter sea ice can block shipping for eight months of the year. By contrast, the interior supports dense forests where summer temperatures reach 95° F (35°C) and frequent lightning storms spawn forest fires, while winter can bring 2–3 meters (6–8 ft) of snow and temperatures below–40°F (–40°C). Southern Labrador is low and full of islands; northern Labrador is a bluff, mountainous coast with peaks over 915 meters (3,000 ft) high rising directly from the sea. Coastal game resources are diverse and plentiful, while on the interior, if you miss the caribou migration, you

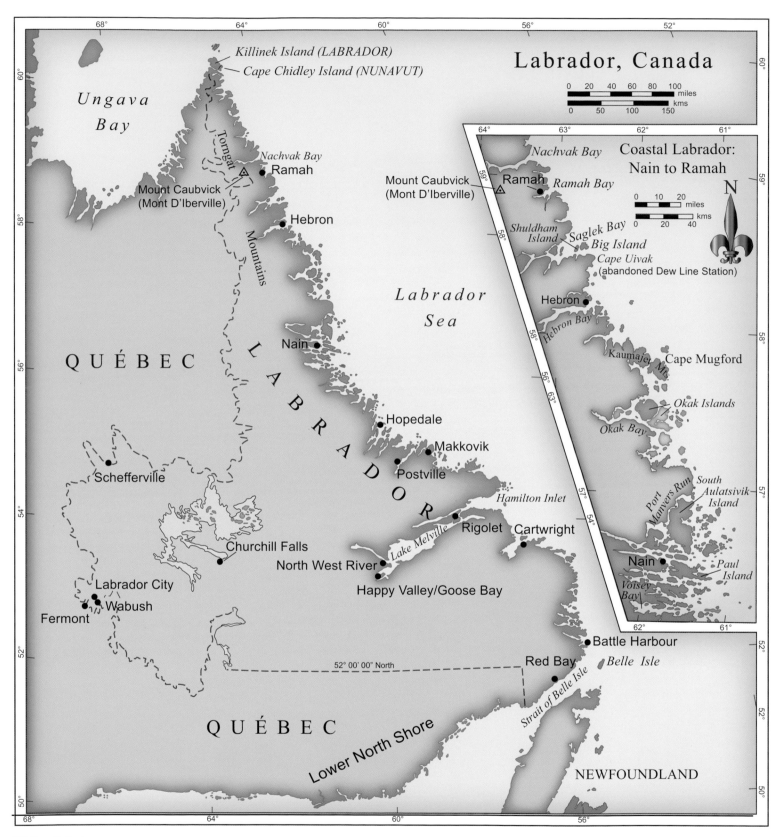

Labrador, Canada

Ungava Bay

Killinek Island (LABRADOR)
Cape Chidley Island (NUNAVUT)

Torngat Mountains

Nachvak Bay
Ramah

Mount Caubvick (Mont D'Iberville)

Hebron

Labrador Sea

QUÉBEC

LABRADOR

Nain

Hopedale

Makkovik

Postville

Schefferville

Hamilton Inlet

Churchill Falls

Lake Melville
Rigolet
Cartwright

North West River

Happy Valley/Goose Bay

Labrador City
Wabush
Fermont

52° 00' 00" North

QUÉBEC

Lower North Shore

Red Bay
Battle Harbour
Belle Isle

Strait of Belle Isle

NEWFOUNDLAND

Coastal Labrador: Nain to Ramah

N

Nachvak Bay

Mount Caubvick (Mont D'Iberville)

Ramah
Ramah Bay

Shuldham Island
Saglek Bay
Big Island
Cape Uivak (abandoned Dew Line Station)

Hebron

Hebron Bay

Kaumajet Mts.
Cape Mugford

Okak Islands

Okak Bay

Port Manvers Run
South Aulatsivik Island

Nain
Paul Island

Voisey Bay

6.02 Map of Labrador
Until the 1940s the population of Labrador was scattered in small coastal villages stretching from Blanc Sablon and the Strait of Belle Isle to Port Burwell on Killinek Island at the entrance to Hudson Strait. Today large population centers have grown up near the former US-Canadian airbase at Happy Valley–Goose Bay and the mining towns of Wabush and Labrador City.

6.03 Nulliak, an Ancient Indian Camp
This summer camp on Labrador's Torngat coast was occupied by Maritime Archaic Indians 4,500 years ago. Its residents mounted expeditions to obtain tool-making stone from the Ramah chert quarries located one day's travel to the north. Nulliak's archaeological remains feature 100 meter long houses and several burial mounds.

can die of starvation any time of year. Contrasts abound at every turn.

Labrador's human settlement patterns also differ widely among regions. The island-shielded central coast provided protection for traditional umiak and kayak travel by the Inuit (Eskimo). To the west, accessed by deep fjords, were the high Labrador barrens filled with small lakes, ponds, and bogs, the home of the Innu (Naskapi) Indians and the huge George River caribou herds (figs. 6.08, 1.08). To the south lay a plethora of islands, barren on the eastern sides facing the icy Labrador Current, and forested on their western shores and in the nearby bays and inlets. From here to the Strait of Belle Isle the proximity of land and sea resources combined with favorable geography and a predominance of offshore winds made the Labrador coast a magnet that has attracted human settlement from all directions.

The coast has been known to Europeans since Viking times and has been the primary settlement corridor of Inuit and Europeans until the modern era. But the interior—the land of the Innu—was not penetrated by outsiders until the mid-1800s and was not mapped even provisionally until after 1900. Hudson's

Bay Company posts and Moravian and Catholic missions served as focal points for Inuit and Innu settlements. The Innu were the only permanent residents of the interior until Goose Bay was established as a military base in 1943 and the mining towns of Labrador City and Wabush attracted settlement in the 1960s. Today most of Labrador's 27,000 people live in these three cities, while the Innu mainly live in small settlements on or near the coast and continue to hunt, trap, and fish in the interior. Combining traditional and modern elements, the cultures of the Inuit and Innu demonstrate a transition between the European-dominated societies of the southern Maritime Far Northeast and the indigenous-dominated ones further north.

THE LAND

Labrador's coast faces the North Atlantic Ocean (technically, the Labrador Sea) and is articulated by numerous fjords, bays, and islands. Its territory extends across nearly 11° of latitude, from L'Anse au Claire at ca. 51° 30'N on the Strait of Belle Isle to Killinek Island at 60° 21'N at the entrance to Hudson Strait. The coast is chilled by the cold Labrador Current, sometimes called "iceberg

6.11 Kameshtashtan Bear Ritual
Like many northern peoples, Innu honor the bear, whose spirit is among the most powerful in their pantheon. Following a communal meal, the bear's skull is usually placed in a tree out of the reach of dogs and other carnivores. Most are skulls of black bear (*Ursus americanus*), but some have been identified as the barren ground grizzly (*Ursus arctos*), which was not thought to have lived east of Hudson Bay but now is believed to have been extirpated in the nineteenth century.

6.12 Shaking Tent Ceremony
This illustration from Henry Youle Hind's *Explorations in the Interior of the Labrador Peninsula* (1863) shows the shaking tent ritual conducted by Innu shamans to cure disease or placate evil spirits. Upon entering the tent the shaman would begin to chant, and soon his struggles with the spirit he had summoned could be seen and heard by those gathered outside.

passed down in the female line were pre-Columbian or a result of the fur trade. Eleanor Leacock (1954, 1969) advocated for the fur trade explanation in one of the earliest examples of feminist anthropology.

The Innu still hold the Labrador interior, its forests, lakes, river, and animals sacred, as they have for thousands of years. While strongly Christian today, they are only a few decades away from the time when traditional religious beliefs governed every detail of their lives. Artifacts from this period include medicine bags containing charms and amulets that helped the Innu connect with spirits that assisted the hunt, attracted fur-bearing animals to traps, and protected people from harmful spirits. The most important rituals, such as the Mokoshan feast involving the preparation and eating of caribou bone grease, were held to honor or influence the master caribou spirit that lived in the Torngat Mountains and controlled men's fate by releasing caribou and other game. With life so dependent on this animal, the Innu were careful to observe the proper rituals and prohibitions, especially after a successful hunt. Most ceremonies were conducted by shamans, religious specialists with knowledge about the ways of the spirit world and whose powers came

6.13 Innu Painted Robe
Innu women fashioned garments of bleached caribou hide painted with intricate multicolored parallel lines and double-scroll motifs. The designs were based on visualizations of men's dreams and were intended to please the caribou spirits. Some believe the design's narrow waist, flaring skirt, and open front may reflect early contacts with European visitors.

sweeping curves that doubled back upon themselves in what is known as the "double-curve" motif. The cut of the garment and similarities to European folk art suggest possible influence, but the content of the designs is wholly indigenous, based on women's interpretation of hunters' dreams. These garments, used in ceremonies associated with the hunt, are among the finest examples of North American Indian art.

THE LABRADOR COAST: EARLY INDIANS AND INUIT

Labrador's coast has a more diverse set of animal resources and a longer and more complex human history than its thinly populated interior (Fitzhugh 1972, 2009, 2013). Camps and burials of its Indian and Inuit cultures are scattered in every river mouth and bay, on every island and headland, especially in the better hunting and fishing locations (figs. 6.34, 6.36–37). The islanded coast with its protected passages and runs was Labrador's most important communication corridor until the era of air flight. Travel on the interior required negotiating physical barriers at almost every turn, and one's payload was limited to what you could carry on your back. Coastal travel required sturdy boats and reasonably good weather, but large numbers of people and gear could be moved quickly over long distances.

Two days' journey north of South Aulatsivik Island, while doing archaeological research for the Smithsonian's Torngat Archaeological Project in 1977, my team made an important discovery that recalibrated everything we believed previously about Labrador's early Indian cultures. During the early 1970s we had excavated a large 4,000-year-old Maritime Archaic Indian settlement at Rattler's Bight on the forested central coast in Groswater Bay, Hamilton Inlet (figs. 6.05–6.07). Our model for reconstructing the long strings of hearths we found was based the single-family tepee-style tents used by the modern Innu. At first we thought these hearth strings were palimpsests resulting from occupations taking place over many years. Then, two hundred miles to the north, at sites in Okak, Nulliak, and Saglek, we found sites of the same culture, but here the hearth strings were contained within low earth-mounded walls that extended 50–100 meters (figs. 6.14–6.15, 6.17). Apparently Maritime Archaic families in both locations were living in longhouses with central hearths spaced 3–4 meters apart. However, only in the stormier north coast were the houses identifiable because their tent walls were reinforced with earth and rock foundations.

from acquiring spirit helpers during fasts, quests, or near-death experiences. The interpretation of dreams was also a prominent feature of Innu spiritual life (fig. 6.12).

Most Innu amulets were small and nondescript, but one item stands out: the painted Naskapi robe (Burnham 1992). The earliest examples, collected around 1700, were already fully elaborated, making its origin uncertain (fig. 6.13). The finest examples come from ca. 1850–1900, when caribou were plentiful and the Innu were engaged actively in the fur trade. These robes, used by the Innu and eastern Cree, are unlike anything created by other northeastern or Subarctic Indians. Rather than being loose-fitting and beaded, they were finely tailored from bleached caribou hide, had narrow waists, flared bottoms, and long sleeves, and were covered with intricate painted designs composed of parallel, multicolored straight lines and

6.14 Nulliak Cove Longhouse
Maritime Archaic sites are usually associated with
forest habitat, but one of the largest sites in Labrador
is located between Hebron and Saglek, well north
of the forest limit. Scores of longhouses like the one
shown here indicate that large populations congre-
gated here over many summers 4,200 years ago to
hunt caribou and acquire stone from the Ramah quarries before returning south to winter in the forests of
the central coast.

6.15 Soapstone Pendants
Excavations at Nulliak have produced unusual pendants,
personal ornaments made of soapstone engraved
with unique geometric designs reminiscent of patterns painted on Innu robes (fig. 6.13) and carved bone
pendants (fig. 4.07) found in Newfoundland on Beothuck
Indian burials. These similarities suggest a 4,000-year
continuity in art and belief.

Although these sites were located in what came to
be known as "Eskimo country," the longhouses turned
out to be from a 4,500-year-old phase of Maritime
Archaic Indian culture, the same tradition that created
the 7,500-year-old burial mounds in southern Labrador at Forteau on the Strait of Belle Isle (fig. 6.10).
Over thousands of years, these Indians had developed
a complex culture focused on a combination of land
and sea hunting: they pursued sea birds, fish, walrus,
seals, and small whales during the summer and caribou and other fur-bearers during winter. Their adaptation was so successful that their culture persisted with
only minor changes more than 4,000 years. During
this time their dwellings expanded from small single-
family huts to longhouse dwellings of 50–100 meters
occupied by twenty to twenty-five families, each with
its own hearth. Their tools included polished stone
axes and gouges made of slate originating in Newfoundland, which they used to make large oceangoing
dugout-log canoes. Chipped stone tools were made
from translucent chert quarried from Ramah Bay, two
days' journey north of the forest boundary between
Okak and Hebron.

6.16 Bishop's Mitre

The most spectacular mountain on the Labrador coast is the Bishop's Mitre, located on the north-facing side of the Kaumajet Range directly across the bay from Nulliak. The mountain is named for its distinctive cleft peak, which resembles the design of a Catholic bishop's hat. Innu and Inuit believed such mountains housed master spirits that controlled humans and animals.

6.17, 6.18 Notched and Cleft Pendants

Many of the soapstone pendants found in the Nulliak longhouses have notched tops or engravings that seem to depict this distinctive mountain on the Nulliak skyline. Perhaps Nulliak people, like the Inuit and Innu, believed the Bishop's Mitre harbored a master spirit and wore the pendant designs as protective charms.

Over time Maritime Archaic mortuary ceremonialism evolved from single burials in low rock and earth mounds to large cemeteries with "red paint-filled" graves like those in Newfoundland and Maine. Throughout these coastal regions groups buried their dead in cemeteries, placed large numbers of stone, bone, birch-bark, and ivory tools with the deceased, and covered the bones with red ocher made from iron-rich hematite mixed with animal fat. Their highly complex, artistic, maritime-adapted culture reached its peak between around 4,500 and 3,500 years ago, when widespread similarities in tools and burial rituals unified the entire region from Maine to Labrador in a single broad cultural tradition. During this period Maritime Archaic people shifted their settlements annually from wintering sites on the central coast to summer sites three hundred miles north on the mountainous Torngat coast, where they hunted caribou and quarried Ramah chert. Their long period of cultural elaboration, coinciding with a period of relatively warm climate, came to a sudden end 3,500 years ago.

The highest achievements of this early Indian history in technology, settlement size, and mortuary ceremonialism, which occurred between 4,000 and 3,500 years ago, paralleled the settlement of Paleo-Eskimos in northern Labrador lands formerly occupied by Maritime Archaic people. For several hundred years both cultures flourished, maintaining separate territorial enclaves within each other's territory for quarrying chert and hunting. Both groups had entirely different histories, beliefs, and physical appearance. The arrival

of Siberian-derived Paleo-Eskimos strangers in traditional Maritime Archaic lands created a remarkable cultural confrontation. Whether this meeting and subsequent interactions were friendly or hostile and whether it contributed to the Maritime Archaic's cultural elaboration and complexity is uncertain. Only the end result is clear: by 3,700 Maritime Archaic culture began to collapse, first in Labrador and a few centuries later in the most isolated island of Newfoundland. Cemeteries were abandoned; the Ramah chert trade ended; and new southern tool materials, tool and dwellings styles, and probably new people appeared. These sudden changes were not one-sided: at the same time Pre-Dorset culture withdrew from the central coast into northernmost Labrador, and a new phase of Indian cultures more similar to the Innu replaced the Maritime Archaic tradition in central and southern Labrador.

From this time forward, the continued presence of Paleo-Eskimos in northern Labrador and Indian cultures on the forested interior and southern coasts ensured that cultural confrontation would be a continuous feature of later Labrador prehistory. The next 3,000 years saw a succession of Indian and Eskimo culture advances and retreats as these groups vied for the prime hunting grounds and lithic resource areas

on the Labrador coast. In warm periods Indian cultures advanced north and reactivated the Ramah chert trade. In cooler periods Indians retreated into the interior forests and southern coasts and Groswater and Dorset Paleo-Eskimo groups advanced south, occupying all of Newfoundland and the northern Gulf. The prime external factor was climate change. Warming meant less sea ice and calmer summer seas suitable for Indian travel on the coast, but caused disruption of caribou herds from forest fires. Cooling resulted in fewer fires, better conditions for caribou and southward extension of sea ice and sea mammals. At these times, Eskimo groups advanced south. During the warm period when the Norse appeared in Greenland, early Innu cultures occupied the Labrador coast as far north as Okak. When later European explorers arrived in the mid-1500s, the Little Ice Age assisted an Inuit expansion into southern Labrador and the Quebec Lower North Shore between 1600 and 1800.

The first Europeans to arrive in Labrador were the Greenland Norse ca. AD 1000. Leif Eriksson was the first European to give Labrador a name, calling it Markland (Forest Land). Sagas recount many Norse adventures in Markland: how Leif's brother Thorvald was killed by a "skraeling" (native) arrow and was buried near a hill

6.20 Hebron Fjord Headwaters
During the Ice Age, tongues of glacier ice cut deep U-shaped valleys as ice rivers descended from the interior to the sea. Inner fjords, like this one in Hebron, have shallow lowlands protected from wind and elements that teem with life, especially char and caribou.

shaped like a ship's keel, and how the Icelandic trader, Karlsefni, arrived with his wife, Gudrid, and their child, Snorri, became the first European born in America.

In 1586 John Davis, an English explorer, scientist, and cartographer, described a visit to an Inuit village on the central coast near Davis Inlet (Clements 1889). At this time Thule culture Inuit were relative newcomers on the central coast, having replaced or absorbed the Dorset Paleo-Eskimo and Innu residents. Davis found the Inuit ravenous for iron and quick to acquire European goods by all means fair and foul; at one point,

under cover of trading festivities, the Inuit attempted to wreck his ship by cutting its anchor cables so that they could salvage its oak timbers and iron. After Davis, history is silent for two hundred years. Basque and Dutch whalers and fishermen visited Labrador sporadically to hunt whales and walrus and trade for Inuit ivory, oil, and baleen (Kupp and Hart 1976). Information is scarce because most of these contacts were ship-based due to the Europeans' fear of attacks on their trading excursions ashore.

Red Bay, located in southern Labrador, saw the appearance of Basque whalers during the 1520s, only a few decades after Columbus arrived in the New World. Between 1545 and 1585, several hundred Basque ships were participating annually in the first energy boom in the Americas—the whale fishery. By 1600 they had decimated the whale stocks and European commercial interest shifted to the cod and harp seal fishery. Trade with the Innu for furs began in the Gulf, and as Inuit settlements advanced south along the Labrador coast between 1450 and 1600 a lively but unpredictable relationship developed as Inuit traders began exchanging whalebone, baleen, ivory, hides, and other sea-mammal marine products from northern Labrador for European boats, iron, weapons, beads, and other goods. By the

early 1700s, assisted by the cool climates of the Little Ice Age, Inuit expanded their settlements into southern Labrador and along the north shore of the Gulf and began to participate in European fishing ventures (see Chapter 5).

One enduring legacy of the coexistence of Inuit and Caucasian settlers is the population of Métis in southern Labrador (Borlase 1994). But a cultural cooperation broader than that between individuals, if it ever existed, was short-lived: by the late eighteenth century combined pressure from Innu and Europeans pushed the Inuit from the Gulf and Strait of Belle Isle back to Cartwright and Hamilton Inlet where their southern boundary remains today. As the Inuit withdrew, French trading posts advanced, reaching Hamilton Inlet in 1743. In that year the French trader Louis Fornel established the first post in Labrador at North West River, near the present town of Happy Valley-Goose Bay at the western end of Lake Melville, the western freshwater segment of Hamilton Inlet. After the Treaty of Paris in 1763, which ended the Seven Years War between England, France, and Spain, Labrador and western Newfoundland territories were awarded to England and settlers like George Cartwright who established posts in southern Labrador and Cartwright. Post servants and Newfoundland homesteaders soon began forming the basis for a Métis society by marrying native women—mostly Inuit—and taking up year-round residence as fishermen, hunters, and fur trappers.

Over time southern Labrador became a European-dominated region with a strong Métis component.

6.25 Our Black Bear
While staging our research out of North West River at the western end of Lake Melville (Hamilton Inlet), we became friends with an old-time trapper named Henry Blake, seen here drying a black bear skin on the wall of his shed. One day when he was ill, he asked us to check the bear trap we had helped him set a few days earlier. We sadly found this young bear in the trap and purchased its pelt from Henry for $20.

< < < OPPOSITE TOP LEFT

6.22 Codfish at Smokey
The small trading post at Smokey on Groswater Bay was a thriving summer fishing depot through the 1900s. Local "livoyers" (slang for "live here") caught and salted their fish for winter, while "stationers" (visiting summer fishermen from Newfoundland) sold their catch for cash. Cod was abundant, and cheap—with fishermen getting only a few cents per pound. But after 400 years of increasingly efficient harvesting and the appearance of offshore draggers, cod disappeared in the late 1970s from Labrador and most of Newfoundland. After a twenty-year moratorium cod stocks are only starting to rebound.

< < < OPPOSITE TOP RIGHT

6.23 Fishing at Rattlers Bight
Local residents Ozzie and Joyce Allen are cleaning their salmon net on a calm day. By 1973 salmon had replaced cod as the cash crop for local inshore fishermen. Plastic nets had replaced tarred cotton twine, making fishing less expensive and more productive, but fishermen still made do with homemade boats.

< < < OPPOSITE LEFT

6.24 Fishing Stage at Rattlers Bight
The dock and storage shed called a fishing stage has been the centerpiece of inshore fishery for centuries. We anchored our skiff off this old stage while excavating one summer at Rattlers Bight. The sound of surf breaking on a shoal (a rattle) would echo across the bight (a broad cove), inspiring the name of Rattlers Bight.

During the heyday of the offshore schooner cod fishery in the late nineteenth century, towns like Forteau, Red Bay, and Battle Harbor were established. Today Battle Harbor's fish flakes for drying cod and its old-time general store are part of a fishing-themed tourist village, but in earlier days it was known to outsiders as the location of Robert E. Peary's 1909 telegraph message announcing his North Pole conquest.

OUTSIDERS IN LABRADOR

Following Louis Jolliet's exploration of southern Labrador in 1694, which documented meetings with Inuit, Labrador began to be explored by scientists and anthropologists. One of the first scientists to visit Labrador was John James Audubon, whose explorations on the Quebec Lower North Shore reached southern Labrador in 1833. His bird studies were followed several decades later by Winfrid Stearns (1884), another ornithologist whose geographical approach produced descriptions of southern Labrador's natural and human history. The first descriptions of Northern Labrador come from geographical explorations in 1894–95 by Albert Peter Low (1896), who traversed northern Labrador and Quebec and discovered the huge iron deposits on the northern interior. Better known to the public were the exploits of Leonidas Hubbard, Dillon Wallace, and their Cree Indian guide George Elson whose ill-fated canoeing expedition of 1903 ended in Hubbard's death. In 1905 a race to Ungava Bay following the same route by rival expeditions—one led by Hubbard's wife, Mina, guided by Elson, and the other

by Wallace—produced publications that attracted a broad following (Wallace 1905, 1907; Hubbard 1908; Davidson and Rugge 1988). Both groups described passing Innu camps for hunting caribou as they descended the George River.

Many Canadians and Americans came to know Labrador from the numerous books written by Labrador's famous itinerant doctor, Wilfred Grenfell (see Chapter 4). Grenfell, who was later knighted by the British Crown for his efforts, arrived in 1892 to operate a floating medical mission for the deep-sea fishers who came to Labrador during the summers from the British Isles, New England, and the Canadian Maritimes. In 1893 Grenfell brought his mission ashore and began traveling to coastal villages by dogsled in winter and, later, in summer on the Grenfell Mission ship *Strathcona*, and that same year established his first hospital at Battle Harbor. That it was better to serve indigenous populations of Innu (Montagnais) and Inuit in the villages where they lived, rather than to move them and their families "to more settled districts," was a central tenet of Grenfell's philosophy as a missionary doctor. For he "believed in Labrador's economic potential and consistently dismissed the notion that the resident population should be shifted to a more agreeable location" (Romkey 2009: 111). His ministry also brought religious and educational services and established worker cooperatives.

Over the next twenty years Grenfell built other hospitals along the Labrador coast and in northern Newfoundland, where he established the mission's

6.26 Inuit Winter Camp
In late winter the Labrador Inuit move to the *sina*, where they camp near the open water lead between the land-fast ice and the moving ocean pack-ice to hunt harp seals and sea birds using flats (small boats) drawn on dog sledges to retrieve the game.

headquarters at St. Anthony. For years Grenfell and his successors expanded the medical services and upgraded its hospitals and nursing stations as part of the International Grenfell Association. In 1991, a century after Grenfell arrived, the IGA sold its entire operation to Newfoundland for $1 and became the Labrador-Grenfell Health Authority, part of the Newfoundland and Labrador government. To this day one still meets individuals from throughout Canada and the United States whose acquaintance with Labrador began during summers as student volunteers, known to IGA staff as "wops" and "wopesses," gender-specific sobriquets for "work without pay." Until the military emerged as an alternate social lifeline to the south, these volunteers and the young doctors who interned for several years at a time provided Labrador's only social connection beyond Newfoundland.

6.27 Mikak and the Moravians
In 1767 Mikak, a young Inuit woman from northern Labrador, was captured by Captain Lucas and taken to England, where she learned English, met the Moravian missionary Jens Haven, and became part of a Moravian plan to Christianize the Inuit. This portrait with her son Tutat was painted by John Russell in 1769. She returned to Labrador and was instrumental in convincing the Inuit to let the Moravians establish missionary stations along the coast.

Considering Labrador's deep history, scholars have been slow to discover Labrador and accounts by historians have been relatively few. One of the earliest is the comprehensive account by William Gosling (1911). Vainö Tanner, a Finnish botanist (1947) produced a two-volume geography of Labrador in 1947. More recent history is covered by William Rompkey (2003). A unique quarterly publication named *Them Days: Stories of Early Labrador* has since 1975 been documenting and preserving the old ways and early days of Labrador. Founded by Doris Saunders of Goose Bay, the journal captures the stories and personalities of "them days" and publishes them in their vernacular diction. Material is drawn equally from Labrador's settlers, Métis, Inuit, and Innu populations, resulting in a history of Labrador cultures and peoples "in their own words." Lynne Fitzhugh compiled these oral-history accounts by native and Euro-Canadian populations into a chronicle of the past (1999).

The early contact-era history of northern Labrador is closely tied to the arrival of Moravian missionaries in Nain in 1771. At the time Inuit populations resided in several winter villages scattered among the outer islands where they hunted (fig. 6.26) seals, walrus, and whales (Taylor 1984; Borlase 1993). The winter land-fast ice in the bays and island passages was prime habitat for ring seals, and the *sina*—the open water lead between the fast ice and the southbound-drifting Arctic pack ice—was where birds and marine mammals could be found all winter. Spring brought walrus and ugruk seals, whose hides provided waterproof boots, foul-weather parkas, and new skins for boats; whale and walrus blubber provided dog food and lamp oil; and walrus ivory was used for harpoons, tools, and ornaments. During summer, villages dispersed to inner-bay hunting and fishing sites. In late summer and fall caribou migrated through the highlands west of the coast, providing food, sinew, antler, fat, and furs for winter clothing. Late fall brought a southbound harp seal migration that filled the caches for winter.

The Moravian Church, originally called *Unitas Fratrum* or Unity of the Brethren, began as a Protestant sect in Herrnhut, Moravia (now Czech Republic), dedicated to piety, teaching, religious conversion, and music. Through connections with the Danish royal family a Moravian mission was established in Greenland in 1744, and in 1752, merchant members of a London congregation outfitted a trading and missionary voyage to Labrador. Since the 1550s hostilities had been a constant feature of the interactions between Inuit and European fishermen in southern Labrador. Encouraged by the British government and imagining they could

6.28 Moravian Mission in Hebron
Hebron, where a Moravian mission was established in 1831, is one of the most productive hunting and fishing regions in Labrador. Although many died during the Spanish influenza epidemic of 1918, the station survived until 1959 when its population was relocated to Nain to facilitate political centralization. Today Hebron's church and missionary complex, prefabricated in Europe, is a major tourist attraction.

be agents of peace, Johann Christian Ehrhardt and six Moravian companions landed at Nisbet's Harbor near Makkovik, built a small house and began trading and proselytizing in the midst of Inuit territory. The mission ended in tragedy when the group, carrying valuable trade goods, was intercepted and murdered during their first attempts of contact with Inuit near Davis Inlet.

Although the Moravian goal was to instill Christianity, the British government's intent in granting the Mission land was to secure the southern fishery from Inuit depredation. They wanted the Inuit pacified through conversion and kept from traveling south where they harassed and raided fishing establishments. In 1765, with government encouragement, Jens Haven, another Moravian with experience in Greenland, initiated a new approach: he made a preliminary agreement allowing Moravians to establish a mission with Inuit his party met in southern Labrador. Then, to ensure commercial traders would be excluded, the Moravians lobbied for an exclusive land grant in Inuit territories in northern Labrador. In 1767 an Inuit woman named Mikak (fig. 6.27) became involved when she was captured by Captain Francis Lucas following an Inuit skirmish at a trading post in Chateau Bay. Lucas began teaching Mikak English and took her to London, where she was introduced to Haven and to London society, becoming something of a celebrity while advancing the Moravian cause. Her portrait was painted by John Russell, and she was befriended by Augusta, dowager Princess of Wales and the mother of King George III, who presented her with royal costumes and jewelry. Returning with Haven and her trousseau to Labrador, Mikak helped convince the Inuit to authorize a mission in Nain. The plan was realized in 1771.

Mikak continued to have an influential role in Labrador Inuit society after her return from England. By the time she died in Nain in 1795, the Labrador Inuit had begun to convert to Christianity, a process that accelerated 1802–5, a period that became known as the religious "Great Awakening" of Labrador. The Mission's acceptance in Nain resulted in its expansion south to Hopedale and Makkovik, and north to Okak and Hebron (fig. 6.28), Ramah, and eventually, in 1904, to Port Burwell on Killinek Island, one of the richest marine-mammal hunting territories in the Eastern Arctic. The mission at Port Burwell closed twenty years later because of costs and difficulty of supplying it, due to its isolation, extreme tides, tortuous currents, and dangerous ice.

In its early years the missions' annual reports to London contained interesting information relating to Inuit demography, culture, and social life (Taylor 1974). But Moravian interest was in saving souls, not ethnography, as seen in the following passage from the report of missionary exploration from Okak to Ungava, which says more about the missionaries than the Inuit.

Our very hearts rejoiced in this place [Hebron], which had but lately been a den of murderers, dedicated, as it were, by the *angekoks*, or sorcerers, to

continues on page 146

Exploring the Torngat Coast

By Wilfred E. Richard

O N MY FIRST TRIP TO NAIN WE CHARTERED a boat captained by Eric Webb. We motored up Port Manvers Run, an inland passage north of Nain, and stopped at a "tilt," a small hunter's cabin near the mouth of Webb Bay, named for our captain's family, who have hunted and trapped here for more than one hundred years. Scattered about were bits of fur and bone, reminders of the annual fall hunt when thousands of caribou sometimes migrate through this region. To the east rose Mt. Thoresby, the highest peak on South Aulatsivik Island.

South Aulatsivik, one of the largest of hundreds of islands scattered along the Labrador coast, is a place of solitude, with snowfields, bears and caribou, and innumerable mosquitoes. It is a land littered with hundreds of thousands of glacial erratics ranging from fist-size to house-size, many perched precariously upon a few tiny rocks. Among this glacial detritus, disorder is legion and the skyline's smooth glaciated curves are punctuated by randomly placed lumpy rocks. These boulders have lain on the surface, immobile, since being dropped by retreating glaciers some 8,000–10,000 years ago.

On my second trip to Nain, I met two more of the Webb brothers: Chesley and Joel, members of a generation of sixteen brothers and sisters with the same Inuit mother. Ches Webb served as captain of the *Viola Dee*, with his brother Joel as first mate and Ches's son, Jerrett, as seaman. The *Viola Dee* cruised east though Harmony Run, named for the ship that annually supplied the Moravian missions on this coast. At the anchorage at Black Island filets of char (*Salvelinus alpinus*) were being dried in the smoke of smoldering blackberry (*Empetrum nigrum*) vegetation and made into *pipsi* (fig. 6.30). Scattered along the shore the remains of Inuit tent rings poked through the thin tundra vegetation. In 2012 Amelia Fay of Memorial University in St. John's, Newfoundland, excavated a sod house foundation near the shore, which she believes may have been occupied in 1776 by the legendary female Labrador Inuk, Mikak, born around 1740 (fig. 6.27).

As *Viola Dee* traveled north, trees, except for an occasional scrubby black spruce, disappeared after Port Manvers (57⁰ North). We viewed icebergs drifting south from northern Baffin Bay and the astonishing uplifted rock strata of the Torngat Mountains with beds of slate, schist, and Ramah chert crisscrossed by giant bands of black basalt. At mission stations, outbuildings, cemeteries, and the remains of the Moravians' rudimentary agricultural efforts persisted in deteriorating greenhouse boxes. We observed animal tracks, types of vegetation and minerals, ate "country food" like Arctic char, black duck, and caribou, and drank from streams. We rarely saw another boat.

In the evenings we read Hans Rollmann's (2002) history of Labrador's Moravian stations. His commentary blended with the stillness of the evening, creating a time for reflection on those early days when Christianity began to infiltrate Labrador Inuit society. Experiencing the same weather and climate as the Moravian and Inuit people within these settlements made history come to life. Of course, what we did not want to come to life was the death and suffering of those living here during the infamous influenza epidemic of 1918. That terrible disease, which was introduced by the *Harmony* during its annual supply voyage, decimated Hebron and annihilated Okak (fig. 6.34).

For two hundred years the Moravian Church, led by German and English missionaries in collaboration with Inuit elders, was the central authority in northern

6.29 Chesley and Joel Webb
Chesley (left) and Joel Webb come from a large Inuit Métis family that became established in Webb Bay north of Nain. Today the Webbs operate charter vessels that take tourists and research teams to the wild, abandoned coasts of northern Labrador, much of which is now incorporated in Canada's Torngat Mountain National Park.

6.30 Drying Arctic char
Arctic char (*Salvelinus alpinus*), which is simply air-dried, not smoked, hangs on the port side of the *Viola Dee*. Known as "Labrador sea trout" until the 1960s, Arctic char, along with caribou, are primary local sources of protein.

6.31 *Harmony* in Soapstone
The annual arrival of the *Harmony* bringing fresh food, new missionaries, and mail, was the highlight of the summer season and the mission's lifeline to England, the Moravian headquarters. Its arrival must have inspired one of the mission's Inuit to carve this representation of the ship with its British ensign, which we discovered on a soapstone outcrop during an archaeological survey in 1977.

Labrador villages. Today Moravian ministers are primarily indigenous, and the church has no political authority. As Vaughan writes (2001 [1994]: 277–278), in the early days Christianity was "seen as an adjunct of civil government, missionaries were often encouraged by the state… Conversion and trade went together; the profits from trade paid for the mission." While the Moravians held high ideals for their mission work and saw themselves as a shield against the evils of the outside world, they could not stem Inuit desire for trade and social contact in the south. Ironically what the Moravians brought to the Inuit beside Christianity and an imposed European social order was disease and paternalistic domination by church leaders who treated the Inuit as their flock of biblical "children." There were other losses too: traditional culture and language, political and economic independence, and cultural identity. Despite these losses, the Inuit of central and northern Labrador survived, and those living south of Hamilton Inlet did not, except as Métis in a European cultural context.

In Nain the young people were curious and easy to talk to, but the community has suffered from the closure of the northern Labrador Inuit settlements and their

6.32 Settlement Ruins
Hebron's Moravian ruins seem haunted, as though the people are out on the land and may return at any moment. On the second floor, slivers of sunlight penetrate the aged roof to shatter the dark. Caribou antlers are a common sight, left by hunters who find the mission site an excellent hunting ground.

6.33 Shuttered Church and Mission Building
The Moravian assembly hall and community house were often combined, but this building also once featured apartments for European missionaries and their families. When we visited in 1980, its windows and chimneys were boarded up to retard deterioration.

consolidation into southern towns, which lacked sufficient game and natural resources to sustain a traditional economy. Although living in a bounteous land, much of north coast's food resources are too distant to be accessed from Nain. Where there is no work for the younger age cohorts, there is restlessness. As a result Nain and other Labrador Inuit and Innu communities suffer high rates of alcoholism, addiction, suicide, and social violence. In this regard the social problems of these populations are similar to those in other regions of the North. Today there is controversy about whether the Inuit benefitted more than they suffered from their European benefactors. Joel was succinct in expressing his thoughts on the diminishing influence of the Moravians: "Free at last!"

THE NORTHERN MISSIONS

Okak, the closest Moravian mission station north of Nain, was established in 1776, five years after Nain. In 1918, after almost fifteen decades of settlement, Okak was ravaged by the Spanish flu, which killed 78 percent of its population. After the epidemic the settlement was abandoned and torched. We found the mission overgrown with head-high herbaceous plants. While searching for tent sites we encountered hidden stone walls and cellar holes and on the beach porcelain plates and tea cups, sea-polished glass, and rusted iron tools. In addition to refuse we discovered (and ate) rhubarb that was thriving ninety years after it had last been tended in 1918.

Of all the abandoned settlements, the most memorable—even haunting—is Hebron, which was established in 1830 and was abandoned in 1959. Located near the mouth of the fjord of the same name, Hebron (figs. 6.28, 6.32–33) is almost invisible from the sea because its buildings have weathered to the same pale grayish-green as the surrounding rocky hills. Although the settlement has been closed for half a century, its church hall, staff apartments, out-buildings, and Inuit dwellings still stand, and abandoned machinery and artifacts of all sorts lie scattered about.

Pockets of gardening soil are rare in Labrador, even more than on Quebec's Lower North Shore and in much of Newfoundland. One of these tiny pockets is found at Hebron, although it was never sufficient by itself to feed the community. Hebron's acres of lush grass, however, are not a natural product; they were created inadvertently by a century of trash and human and animal fertilizer. Despite access to a rich marine life and abundant caribou, Hebron and the other Moravian missions had to be supplied from Europe.

Only five years had passed since my first visit in 2000, but in that short period Hebron's historic buildings had deteriorated markedly. The assembly hall roof leaked and water was pooled on the second floor. The front door had to be reinforced with boards to prevent collapse. The Inuit Labrador Association, now the Territorial Government of Nunatsiavut, had begun repairs the previous year in order to preserve the mission architecture as an historical treasure. Hebron's buildings were transported from Germany in the 1830s as prefabricated components, each timber carefully numbered with Roman numerals for installation. Restoration is now almost complete.

North of Hebron, even shrubs vanish, and moss and lichen begin to dominate the land. The range of the bakeapple extends from northern Newfoundland and the Quebec Lower North Shore to the furthest northern reaches of Labrador. On the central Labrador coast one also encounters a flying insect that looks like a bee but is a legendary fly known locally as a "stout." This robust creature takes a chunk of flesh rather than a sip of blood. Yet even as the land became harsher and more rugged, the presence of humans continued. Undaunted by the towering Torngats were numerous inuksuit, the human-like stack of stones with a name that means "like a person." These Inuit monuments mark hunting sites and locations in the caribou drive, dangers, directions, spirit places, or, as Norman Hallendy surmises, a companion to an Inuk in a lonely land (Hallendy 2000). Other cultural "footprints" include stone grave cairns, food caches, and innumerable

6.34 Ancient Inuit Village
Near the Moravian Mission station in Okak are the remains of an Inuit settlement that in the sixteenth to eighteenth centuries contained numerous large multifamily communal houses. Three or four extended families lived together in each of these dwellings under the authority of a powerful leader known for his prowess as a hunter and for being effective at trading with Europeans.

6.35 Ramah Chert Deposit
The two snowfields and talus slopes near the horizon mark the location of a surface mine of Ramah chert. An *inuksuk,* visible from the sea (lower right corner of photograph), points the way to this precious lithic resource utilized by prehistoric peoples in Labrador for more than 6,000 years.

stone tent rings of former Inuit and Paleo-Eskimo camps (fig. 6.36–37). Even those dating to the Thule and Dorset cultures show the nutrient effect of fertilization from organic materials. The human footprint has not vanished with its creators—it continues to grow.

North of Hebron we cruised past the wide mouth of Saglek Bay and entered the most forbidding part of Torngat's lair, the region around Ramah Bay where the powerful Inuit spirit, Torngarsoak, known to control caribou movements, resided. This area is also the site of another Moravian station (1871–1904), but to archaeologists it is a sacred site of a different sort—the location where Indian and Eskimo peoples obtained Ramah chert (fig. 6.35) for making tools and weapon points for thousands of years before the appearance of European iron.

Situated just short of North 59°, Ramah Bay is in the heart of the Torngat Mountains and is part of the newly designated Torngat Mountains National Park Reserve. The Torngats are the defining geological feature of this region, and its rocks are known to geologists as the Ramah Series, a layered series of Paleozoic strata cut by numerous bands of black basalt dykes (fig. 1.17). The Ramah Series contains one of the world's greatest visible concentration of dykes, which are produced when molten magma is intruded into cracks and fissures deep in the earth's crust. When the zircon crystals in these dykes were dated in the 1970s at 1.5 billion years they were the oldest rocks on earth. (Since

then they have been outclassed by much older rocks from Hudson Bay, Greenland, and South Africa that date to ca. 2.5–3.8 billion years.)

In Ramah, we set up camp next to the old Moravian mission building foundation. The nearby waterfall that supplied the mission filled our water buckets and served as a frigid but refreshing shower. Because of the threat of polar bears and black bears we had erected an electric fence around our sleeping area. Whether or not such a device would deter bears is questionable, but it provided psychological comfort. While exploring the surrounding hills we were rewarded with numerous sightings of polar bears, black bears, and caribou. One day we decided to bathe in two groups: first women, then men. During the fifteen minutes while the genders were shifting places, a polar bear footprint appeared in the soft sand next to the waterfall. This incident reminded us of the danger of traveling in these latitudes without a rifle.

Warming climates and the associated reduction in polar sea ice has resulted in polar bears increasingly turning to land as a base for hunting and foraging. For reasons biologists have yet to determine (but probably because of plentiful seals and caribou), large numbers of polar bears are being reported along the Torngat coast. Over the long term polar bears are not likely to thrive on land due to their evolutionary specialization as ice-based seal hunters. Today when they come to land they sometimes encounter

black bears, which in recent years have been spreading north into these same tundra regions, and they too are taking caribou.

The current situation is different than in the past, when grizzly bears also occupied the Quebec-Labrador peninsula. For years, biologists had speculated on the possibility of a Labrador grizzly based on Innu and trapper reports of a reddish-colored black bear. The issue was resolved when a bear skull from an historic-period Inuit site in Okak was identified as a grizzly, otherwise known as the Arctic brown bear (*Ursus arctos*) (Spiess and Cox 1976; Loring and Spiess 2007). This young adult was smaller than

the barren ground grizzlies known from Hudson Bay to Alaska. Biologists believe that the Okak grizzly was among the last of a population that became isolated from the more robust western population whose habitat included an important prey species absent in Quebec-Labrador: the Arctic ground squirrel, known in the west as sik-sik (*Spermophilos parryii*). With its population stressed and its physical size diminishing, the likely cause for its extirpation was the eighteenth- and nineteenth-century expansion of the fur trade and acquisition of firearms by the Inuit and Innu. The discovery that grizzlies had occupied the northern Labrador-Quebec peninsula until recently helps explain the expansion of black bears into the forest fringe and tundra and the occasional inclusion of caribou as a part of their diet. For a region designated as a national park, the polar bear influx has created a challenge for wildlife biologists and park officials, as well as for wilderness hikers and kayakers who, according to park rules, may not carry firearms. To ensure their safety, regulations require that they travel with licensed guides, who are mostly Inuit and so may be armed.

On our treks around Ramah Bay, we encountered large numbers of caribou, which, like the solitary polar bear, were curious but not intimidated. Char were also plentiful, having rebounded from overfishing in the 1970s when a fleet of small collector boats run by a fishery cooperative in Nain called regularly at Inuit camps from Nain to Nachvak to pick up salted char. In these years the Newfoundland government subsidized local fisheries and even started a longliner construction business in Postville to replace the aging collector vessels. This motivated many of the north-coast Inuit whose villages had been closed to return during the summer to their old haunts to hunt and fish. Fleets of thirty-foot Newfoundland trapboats had replaced the traditional Inuit skin-covered umiaks. More durable and equally seaworthy, the trap boat evolved from the Basque whaling shallop, which was sailed and rowed; the trap boat also had originally been sailed, but by the mid-nineteenth century was powered by a single-cylinder hand-cranked Acadia gas engine that, once running, emitted an ear-splitting, rapid-pace "bam-bam-bam-bam…" that could be heard for miles. Like Newfoundlanders, Inuit used them for tending cod-traps, but also for their summer travels to the north coast char fishing and caribou-hunting camps. By the late 1970s this seasonal migration came to a close as elders passed on, the char population collapsed, and government support for regional Inuit co-ops declined.

Everywhere we went we found the ground littered with flakes and broken tools of Ramah chert, the raw material of choice for almost all of Labrador's early cultures. At

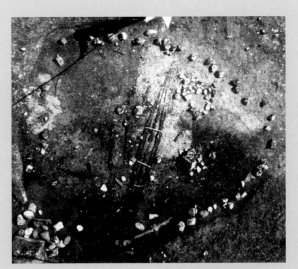

6.38 Ramah Chert Tools
Translucent stone marked with black stripes and pyrites crystals are diagnostic features of chert originating from the ancient rocks of the Ramah series. Ramah chert beds outcrop intermittently over a distance of fifty kilometers, from the northern side of Saglek Bay to Nachvak. Ramah chert has been traded as a raw material and in finished tools across cultural and geographic boundaries for thousands of years.

Hilda's Creek at the northeastern entrance of Ramah Bay, the translucent, sharp-edged detritus (fig. 6.38) formed a deep deposit left over from ancient quarrying and tool production. Following the trail of debitage (the by-products of tool flaking) upslope, we soon found ourselves on a trail marked by a stone *inuksuk*. Passing a pool in which red ocher had accumulated from the natural weathering of iron-rich rocks, the trail rose into a broad, bowl-shaped hollow several hundred meters wide with snowfields and rock scree clinging to its steep walls (fig. 6.35). This bowl, known geologically as a cirque, had been carved out by a small glacierette during the Ice Age, and as the snow turned to ice and moved downslope, it plucked out chunks of rock until it had eaten down to the bed containing the meter-thick layer of Ramah chert. Geologists disagree whether this translucent rock is the product of a solution of liquid quartz (making it a chert) or a metamorphized bed of quartz sand grains (making it quartzite)—the end-products can look indistinguishable. In either case the result was a lithic material of uncommon distinction, translucent with beautiful streaks of black manganese pigment and small rusty crystals of iron pyrites. At this point in its history the freezing and thawing action of water seeping into the chert bed had opened fissures and released chunks that could be pried out and flaked into tools. Prehistoric Indian and Eskimo peoples had been mining this outcrop for thousands of years and trading Ramah chert artifacts as far south as New England, Maryland, and Michigan (Loring 2002).

the service of the devil, to hear the cheerful voices of convertedheathen, most melodiously sounding forth the praises of God, and giving glory to the name of Jesus their Redeemer. Peace, and cheerful countenances dwelt in the tents of the believing Esquimaux. (Kohlmeister and Kmoch 1814: 16)

Economic success followed the early years of financial losses when the Moravians overcame their aversion to worldly matters and opened mission stores to provide staples, excluding firearms and liquor. Their economic policy was intended to stem the continuing tradition of southern voyaging, which often resulted in drunkenness, Inuit and European deaths, and introduction of European diseases. The Moravians soon had to revise their gun policy because rifles had by this time become an important element of Inuit subsistence. Eventually the Inuit shifted to a combination of more localized subsistence hunting, fishing, and trapping, and their conflicts with Europeans and Innu receded.

Lucien Turner's ethnography of the Inuit residing on the Ungava Bay coast in 1882 provides better information on traditional Inuit culture and religious practices than that of the Moravian missionaries, even though it was based on observations from seventy years later. The Inuit of Ungava Bay were closely related to the Labrador Inuit and had been involved with Hudson's Bay Company traders for several decades, but their religious beliefs were more intact because they remained outside the Moravian sphere (Loring 2009). Inuit cosmology saw the world as being populated mostly by malevolent spirits of which Torngarsoak was paramount. While chiefs presided over worldly matters, shamans—assisted by helping spirits—interceded between people and the master spirits to secure game, cure disease, or bring harm to competitors. As Turner observed:

The implicit belief in these personages is wonderful. Almost every person who can do anything not fully understood by others has more or less reputation as a shaman. Some men, by observation, become skilled in weather lore, and get a great reputation for supernatural knowledge of the future weather. Others again are famous for suggesting charms to insure success in hunting, and, in fact, the occasions for consulting the conjurer are practically innumerable. One special qualification of a good shaman is the ability to attract large numbers of deer or other game into the region where he and his friends are hunting (Turner 2001: 196).

The Inuit celebrated special events or times of the

6.39 Innu Woman and Child
Maintaining the tradition of passing wisdom down from elders to grandchildren has become a focus of Innu cultural education at camps like Kameshtashtin in the interior or northern Labrador.

year with festivals, most of which occurred in fall and winter when people were in their winter villages and their food caches were full. A good caribou hunt or the killing of a large whale would be followed by days of dancing, singing, and gift-giving. Moravians saw these festivals, which were often presided over by shamans and included "lights-out" adult activities—as being the work of the devil, and fought them at every turn. After several decades the Inuit living around the Hopedale, Nain, and Okak mission began to accept Christianity, but the Moravians' struggle against the heathen north coast shamans continued throughout the nineteenth century.

CARIBOU AND CULTURE CYCLES

Their constant migrations and the fact that the forest fringe was poor habitat for fur-bearers made the northern Innu (Naskapi) indifferent trappers. On the other hand, they were the world's most dedicated caribou hunters and for this reason have been seen by anthropologists as models for gaining insight into societies of ancient Ice Age reindeer-hunters of northern Europe and the earliest Paleo-Indians of North America (Loring 1997).

After 1900 the George River caribou herd began a precipitous decline that made their migrations unpredictable, and the situation for the Innu changed drastically. Speculation on the cause of this decline has ranged from natural cycles and warming climates to increased predation by wolves. Human intervention may also have been a factor. The combination of large caribou herds and the presence of trading posts that supplied firearms, as well as staples, had encouraged the Innu and Montagnais to move north of their normal forest territory in order to access caribou in the northern

barrens. Human populations had probably grown beyond a sustainable level, and when the herds began to diminish and their migration routes changed, the Innu often found themselves stranded, too weak to relocate. William Brooks Cabot, a gentleman-adventurer from Boston whose idea of a summer vacation was to mount solo canoe trips (1903–1910) in search of the exotic Naskapi found them thriving (fig. 6.09: Cabot 1912, 1920; Loring 1995):

> My objective was Indians. They were people in the primitive hunter stage. Nowhere else, perhaps, was the like of these Indians to be found, a little group of a race high in personality, yet living substantially in the pre-Columbian age....They lived under their own law, in their old faith unchanged (Cabot 1912, quoted in Dillon 2001).

During these years and later, reports from the Moravian missions and trading posts in Labrador and Ungava record years of hardship, including starvation of whole groups of Innu. It was not uncommon for refugees to appear at the coastal posts and missions looking for supplies to take back to their starving families. In the decades following the 1920s, the Ungava and Labrador caribou populations continued to drop, reaching a low of 5,500 in the 1950s before increasing to a peak of over 800,000 in the 1990s. Since then the herd began a free-fall to about 25,000 in 2012, with no sign of resurgence. Once again, overgrazing and wolf predation are cited as primary factors (Bergerud et al. 2008). Biologists now

6.40 Children Posing, Nain
Many Inuit residents of Nain have been displaced by government fiat from their original home locations in more remote regions of northern Labrador. Sometimes these relocations happened in the aftermath of epidemics, but often it was in the interest of providing, more efficiently, centralized government services.

7.01 Bordering Eclipse Sound
At the southeastern tip of Bylot Island,
facing Baffin Bay, Button Point has been
occupied by Inuit hunters for thousands
of years. Today's Inuit from Pond Inlet
maintain a Hunters and Trappers station
here beneath the towering mountains of
Baffin Island seen across Eclipse Sound.

7 Nunavut – Inuit, Whalers, and Explorers

The whole complex Arctic hunting society was, I believe, meticulously formed by trial and error over many thousands of years.

James Houston, *Confessions of an Igloo Dweller*

SIXTY NAUTICAL MILES NORTH OF Killinek Island, across the tumultuous waters of Hudson Strait, lies the southern tip of Baffin Island, known in Norse sagas as Helluland or "Slab Rock Land." The Norse were perceptive geographers whose naming traditions accurately described the economic features of the Maritime Far Northeast—at least from a Norse perspective. Southwestern Greenland was a suitable pasture land for livestock; Helluland was a high rocky land of little use to farmers. Markland, "Forest Land," encompassing central and southern Labrador, was a land that could supply timber for building boats and houses, and Vinland was a rich land where wild grapes and grass held promise for Norse settlements similar to those in their Scandinavian homelands. To the Norse, northern Labrador and Baffin, with their mountainous coasts and deep fjords, were indistinguishable. Today, geographers, botanists, geologists, and even anthropologists would agree; the distinctions between the Torngat coast of northern Labrador and Baffin Island are largely due to recent political history. In fact, the Inuktitut term for the islands around Killinek, *taujat* or "stepping stones," embodies Inuit recognition of their geographic and cultural continuities and shared history.

Today Baffin Island, which stretches 1600 km (1,000 mi) from Hudson Strait to Lancaster Sound, is part of the newly created Canadian political entity called the Federal Territory of Nunavut (fig. 7.02). Nunavut was created from the earlier Northwest

153

7.04 Indigenous Building Materials
This Thule site at Radstock Bay, Devon Island, contains enormous whale bones, which were used to construct a winter house. Most of the bones at this site are from young whales, which were easier to kill and drag home than adults.

Pre-Dorset people probably looked like the Inuit of today, both having ancestry in northern Asia. The Arctic Mongolid physical type shows adaptation to cold Arctic climate of northern Asia: high cheek-bones, a broad flat face, and a robust torso with muscular limbs; these were accompanied by a shortening of the nose, torso, limbs, and fingers, producing a compactness of body that facilitates heat retention. Pre-Dorset people made small triangular points that were fitted to arrows shot from powerful Asian bows. Bow and arrows gave them a military advantage over other Native American cultures that used only heavy spears and hand-thrown darts; for some reason the bow and arrow was not adopted by American Indians for another 2,000 years.

When Pre-Dorset people entered the eastern Arctic, the land was unoccupied, but game was plentiful. Pre-Dorset people were efficient caribou and musk-ox hunters but had only a rudimentary technology for hunting sea mammals, which required toggling harpoons. It seems likely that they learned about harpoons from the Maritime Archaic Indians of Labrador, for by 3,500 years ago, harpoons similar to those used in Labrador are found in Pre-Dorset sites, and seals and walrus begin to become important prey. Pre-Dorset people had a highly mobile life, intercepting game throughout a band's territory, using only tents both winter and summer, heated with fires fueled with driftwood and greasy animal bones. Despite a seminomadic life, they produced remarkable bone and ivory art, and their chipped stone tools made from carefully chosen colorful cherts and quartz crystal have a jewel-like artistry. They probably believed that making their tools and weapons beautiful showed their respect for the animal spirits and deities that controlled their destiny.

Around AD 1250–1300 a new cultural tradition with roots also in Siberia and Alaska arrived in the eastern Arctic. This Neo-Eskimo or Thule tradition culture is named for an archaeological site at Thule, Greenland, on the opposite side of the Arctic from Alaska. Thule people had learned to hunt large whales around Bering Strait, and with the warm climate of the Medieval Optimum, which opened the ice-filled channels among the Canadian Arctic islands, Thule whale-hunters moved swiftly into the Canadian Arctic and northern Greenland. By this time the Norse had established colonies in Greenland, and the desire to obtain iron from the Cape York meteor fall or by trade, for the Norse may have provided added incentive to migration. Thule was a large and boisterous,

rest of northern Canada and Greenland are found in abundance around Eclipse Sound (fig. 7.01), the east-west waterway between Baffin and Bylot Islands, which constitutes a gateway, like the openings of Hudson Strait and the Gulf of St. Lawrence, but this one leads to the western Arctic via the real Northwest Passage. The Bylot Island region has many Dorset and Thule archaeological sites, including Button Point (fig. 7.01) and Karsuq, "Place of Many Rocks," a Thule site a few kilometers west of Button Point, with a profusion of Thule stone dwellings, human graves, bone middens, and raised beaches.

Each of the major cultural shifts—from Pre-Dorset caribou and musk-ox hunters (4,200–2,500 years ago), to Dorset walrus and seal hunters (2,500–800 years ago), to Thule Inuit whalers (800–400 years ago)—corresponds to a climatic change. By 4,500 years ago glacial ice had disappeared from the central Canadian Arctic, and warm climates and ice-free summer waters helped Paleo-Eskimos expand from Alaska into Canada, Greenland, and northern Labrador, where they became known as Pre-Dorset. Over 2,000 years the Pre-Dorset evolved into the Late Paleo-Eskimo Dorset culture during a cold period that required new sea ice hunting methods. Thule whale hunters spread into northern Canada in AD 1250–1300 during a warm phase, and with the return of cold weather ca. 1400 Thule people in the central Canadian Arctic had to abandon whaling and shift back to a walrus- and seal-hunting economy like that practiced by Dorset people. In this way Thule was transformed into Inuit culture of the historical period.

7.05 Thule Winter House, Resolute, Cornwallis Island

A passageway carved out of the tundra kept cold air below the living level of this Thule winter house. This early version of climate control is one adaptation that allowed the Thule people to survive in a hostile climate.

competitive culture, equipped with sled dogs, large skin-covered whaling boats, and powerful sinew-backed bows. The resident Dorset people were no match for this onslaught, and within a century or two most were driven into marginal areas, unattractive to Thule whalers, or had been killed or absorbed into Thule culture. Thule's use of Dorset harpoon styles, soapstone vessels, and igloos indicates a degree of contact and acculturation. Genetic evidence of Dorset–Thule interaction has been limited by the scarcity of Dorset human remains.

Thule culture spread quickly throughout the central Canadian Arctic and into northern Greenland and northern Labrador. In most of these areas they reoccupied abandoned Dorset dwellings (fig. 7.05), making houses that were larger and better insulated. Unlike the Eskimo tradition, Thule people did not make chipped stone tools, relying instead on tools of bone, ivory, and slate. Similar slate tools had been used in Alaska, and this technology produced sharp edges like those on iron tools.

DEVON ISLAND

Shaped like an L lying on its side in a roughly northwest direction, Devon Island extends through 20 degrees of latitude. Devon Island, on the north side of Lancaster Sound, the real—rather than

fabled—Northwest Passage, has a large number of prehistoric Thule Eskimo settlements containing houses made of sod, turf, and whale bones, and food caches and burial cairns made of boulders. The earliest of these sites dates to about AD 1250, the time of the initial Thule migration when the Canadian Arctic island channels were ice-free enough to allow whales to pass from the Pacific to the Atlantic. The latest sites were abandoned during a cold period known as the Little Ice Age, which lasted from about AD 1450 to 1750. During this period Lancaster Sound and the passages to the west became ice-bound throughout the year, restricting Thule access to whales, driftwood, walrus, seals, and other resources crucial to their survival. These same climatic conditions had brought calamity earlier to the Norse agricultural settlements in Greenland. By 1500 the High Arctic was abandoned and with a few exceptions remains without an indigenous population to this day. Devon Island, at 55,247 km^2 (21,331 mi^2), is almost three times the size of the Canadian Province of Prince Edward Island and four times the size of the American state of Connecticut; it is the world's largest uninhabited island.

The bones of walrus, whale, polar bear, and caribou found at these archaeological sites testify to the skill of Thule hunters (fig. 7.04). Harpoons

7.06 Kuyait Site, Frobisher Bay
In the early 1990s a Smithsonian-Canadian team investigating the 1576–78 voyages of Martin Frobisher found Elizabethian-period artifacts at Inuit sites including Kuyait, only a few miles from Frobisher's mines.

that toggled beneath the skin and blubber of large sea mammals had been used earlier by Dorset hunters in the Eastern Arctic, but hunting large whales required the use of inflated seal-skin floats and large skin boats with well-trained crews directed by experienced whaling captains. Social organization was as integral to success in whale hunting as the hunting equipment itself. During the twelfth to fourteenth centuries Basque people independently pioneered the hunting of large whales in the Bay of Biscay, but they used primitive barbed harpoons and did not utilize floats, so they had to remain physically attached to the whale rather than letting it expend its energy fighting the floats. Although Thule people were in contact with European whalers shortly after 1500, Europeans did not adopt Inuit technology until the seventeenth century, when the Dutch began whaling with Basque and Inuit collaboration in Labrador and Davis Strait.

Cornwallis Island (Resolute)

West of Devon Island, across Wellington Channel, is Cornwallis Island and the Inuit/Canadian village of Resolute, the modern administrative hub of the High Arctic. The Inuktituk name for Resolute is Qaqsuittuq, which translates as "place with no dawn," an apropos term considering its location eight degrees above the Arctic Circle. In 2006 Resolute had a population of 229. Resolute was an important American air base in World War II, and it continues as a communications and transportation center for the Canadian High Arctic.

Resolute has a large prehistoric Thule Eskimo site, which was first excavated by Henry Collins of the Smithsonian Institution in the early 1950s and produced some of the first iron and ceramics found in the eastern Arctic. Today several of its underground houses, with roofs supported by whale mandibles and ribs, have been reconstructed. Thule people utilized an ingenious method of keeping their houses warm in the winter. First they excavated about one meter into the ground; then they constructed a sod-covered subsurface entry passage that had at its inner end a large vertical slab that acted as a "cold trap" to prevent cold air from entering the house. One had to step over the stone slab and up onto a higher-level floor that served as the family's working space. Finally, they built a sleeping platform at a still higher level where all the warmest air in the

7.07 Igloolik Bowhead Hunt
Alaska's Thule Eskimos brought the tradition of hunting large whales into Canada and Greenland around 1250–1300. The large quantity of meat, blubber, baleen, and bones supported larger populations and fed the teams of dogs that were crucial to the Inuit way of life. Although dogs are fewer today, whale hunting continues as an important source of nutrition and social and cultural identity. Today's bowhead stocks are carefully monitored, and in recent decades their population has been increasing.

7.08 Sharing *Maktaaq*
Inuit consider the inner skin and outer layer of blubber of white whales and bowheads a prized delicacy, rich in nutrition and vitamin C. Here two elders enjoy *maktaaq* after a successful bowhead whale hunt in Igloolik in 2002.

dome-shaped room collected. A small vent hole in the roof could be opened or closed to regulate air quality and temperature.

Similar houses continued to be used by later Inuit people throughout the North American Arctic and Greenland into the mid-twentieth century. In Alaska, Thule and later Eskimo houses usually were rectangular because they were made with driftwood timbers. In the eastern Arctic they were built with whale ribs and mandibles and were round and sometimes had multiple lobes for extra rooms. After the

decline of whaling and with the appearance of iron tools, Inuit houses began to be built with driftwood or timbers and became rectangular, and in the eighteenth century in Greenland and Labrador, these structures were large enough to house extended families of thirty to forty people. Inuit territories from Labrador to the Canadian Arctic and Greenland contain thousands of these winter dwellings, each type representing a different chronological period or regional variant.

For more than 4,000 years the Inuit lived successfully in the Arctic with a small number of core technologies: tailored fur clothing, soapstone oil lamps and pots, snow goggles, harpoons, snow houses, and skin boats. However, one item, above all, is responsible for the development of the large, vibrant society that came with the Thule and still remains crucial to modern Inuit: the sledge or *komatik*. Ice travel in Nunavut involves riding in a *komatik* (fig. 7.09) that today is usually pulled by a snowmobile rather than a dog team.

The *komatik* is about a meter wide and seven meters long and has a box-like cabin amidship for passengers and gear. These sleds have a structural resemblance to Inuit kayaks and *umiaks*, the latter being large skin-covered boat used both for transport and whaling. These conveyances are always fastened using thong or line lashings, rather than nails or screws, in order to provide flexibility in waves or over rough ice, without which the wood frames of these

159

7.09 *Komatik* Construction
Flexibility is built into the sled, allowing the *komatik* runners and cabin to bend around obstacles, where more rigid assembly would break up on rough ice. The same technique is used in constructing kayaks, which must flex when passing over waves.

crafts would shatter. The Canadian Inuit *komatik* differs from the Greenlandic dog-pulled *qamutit*, which are shorter and lack a cabin, and the *qamutit* also has two vertical posts rising from the rear ends of the runners, which allows the driver to push and maneuver the sled over rough ice and reflects the Alaskan heritage of the original Thule migration.

Whether crossing flat ice, open leads, or pressure ridges, sledges, *komatiks*, and snowmobiles all operate more or less the same way. The biggest risk in spring travel is crossing open leads or crevasses that are more than two meters wide.

7.10, 7.11 Seal Hunting
Guide David Suqslak unsuccessfully thrusts his harpoon into a ringed seal's breathing hole, where his son Curtis later succeeded with a rifle. Seal—particularly ringed seal—is a primary food of Inuit and is traditionally used for clothing, lamp oil, and many other products. Curtis displays his take, most of which we ate on the ice—both raw and cooked.

ADVENTURES ON THE FLOE EDGE AROUND BYLOT ISLAND

By Wilfred E. Richard

7.12 Sea Gooseberry
This minute jellyfish, the incandescent sea gooseberry (*Pleuro brachiapileus*), is part of an almost complete spectrum of life that returns to the floe edge.

THE PORTION OF NUNAVUT WITH WHICH I AM most familiar is the entrance to the fabled Northwest Passage (fig. 7.03), where Baffin Bay merges with Lancaster Sound, north of Bylot Island. My visits began by ship in 2002, followed by five expeditions to the Baffin Bay floe edge, that is, the open water between the drifting sea ice and the ice that is frozen fast to the land, between 2003 and 2008. My objective for these trips was to capture the activity of wildlife as it returns to the waters of Lancaster Sound and Baffin Bay in late May through late June. Depending on winds and tides, the floe edge, known to Inuit as the *sina*, may be a wide, open lead or a choked corridor filled with different types of ice such as grease ice (slush-like ice), hardened pressure-ridge ice, or parts of ancient icebergs from Greenland or from as far away as Siberia.

Surprisingly, sea ice is not just frozen water; rather, it is an ecosystem that sustains a complex food web beginning with algae, phytoplankton and zooplankton reaching up to fish, birds, and vertebrates like foxes, polar bears, sea mammals, and humans (Campbell 1992). Barcott (2011: 31) compares this Arctic food web to that of a farm: "ice and snow are to the Arctic what soil and rain are to the temperate latitudes." This profusion of life in a seemingly desolate, icy setting is fueled by the long hours of summer sunlight, which results in huge blooms of phytoplankton.

Here the greening that signals the onset of spring in the south occurs under the sea ice. As day lengthens in the Arctic, marine phytoplankton form the basis from which algae grow, feeding jellyfish (fig. 7.12), and so on up the food chain. Only recently have scientists discovered that the floe edge has one of the most complicated and concentrated food webs of any environment on earth.

Besides being a habitat for seals, walrus, and whales, the floe edge also provides a place for rest, birth, and sanctuary for the region's master predator, the polar bear. Polar bears use the sea ice as a hunting platform, and with the reduction of Arctic pack ice, polar bears have begun to explore habitats on land. Land sightings are increasing, and several DNA-confirmed offspring of polar bear and grizzly matings have recently been documented (Revkin 2012; Gorman 2012). But *Ursus maritimus* is a marine creature and consequently may be deficient in land hunting skills. Recent anecdotal evidence suggests that a polar bear on sea gains weight while a polar bear on land losses weight. Debate continues about whether polar bears will develop a seasonal terrestrial adaptation. In short, melting sea ice means more than a change in scenery for humans and animals; it may eliminate or change an entire system of life that has evolved to sustain itself in a cold climate.

When I arrived in 2002 at Pond Inlet, a village at northernmost Baffin Island, which is the easiest point of departure to Sirmilik National Park, a magnificent iceberg was stranded near the settlement; a year later, this berg still dominated the sky and dwarfed villagers loading their *komatiks* for the seal hunt out on the ice. Pond Inlet's Inuit name is Mittimatalik, meaning "where Mittima is buried," although no one has ever been able to find out who Mittima was (*Nunavut Handbook* 1998: 323).

Sirmilik, which means "The Place of Glaciers," is the third largest of Canada's national parks and encompasses parts of Baffin Island and all of Bylot, which is covered with glaciers, comprising about 3 percent of Canada's glacial ice (Dowdeswell and Hambrey 2002: 95). Its surrounding

7.13 Polar Bear and Tent
This bear investigated a vacated camp on the floe edge. Polar bears' thermodynamic qualities include the insulating effect of hollow hair and dark skin to better absorb solar radiation.

7.14 Mother and Cubs
Polar bears usually give birth to two cubs per litter, either in dens in the pack ice or on snow-covered slopes of Arctic islands.

waterways—Baffin Bay, Eclipse Sound, Navy Board Inlet, and Lancaster Sound—are a nursery for fish, sea mammals, and migratory birds. In the Arctic of Canada and Greenland, this "icescape" is an ecosystem that provides a suite of services—a home for seals, walruses, and polar bears; a source of protection for these and other wildlife; a highway by which human populations access relatives and friends; and a fresh meat larder (Eicken 2010: 360). *Sila*, sea ice, connects people, with their environment, and their weather (Holm 2010: 147).

7.15 Polar Bear Paw Print
A glove alongside a paw print in slush conveys the size of these massive, burly animals.

According to Parks Canada (2003), Bylot Island contains the most diverse bird population in Canada north of 70 degrees latitude. Seventy-four species of birds are found among the Baillarge Bay seabird colony, the Bylot Island Migratory Bird Sanctuary, Oliver Sound, and Borden Peninsula. When the sea ice melts in late May and into June, 75 percent of the world's population of snow geese (*Chen caerulescens*), of which there are now millions, nest on Bylot Island.

Richard Sale's *A Complete Guide to Arctic Wildlife* (Sale 2006) identifies a plethora of bird species found in the Arctic, many of which annually migrate between the Arctic and Maine (Fig. 5.13) Those most common around Baffin Bay are long-tailed jaeger (*Stercorarius longicaudus*), snow bunting (*Plectrophenax nivalis*) (fig. 7.19), northern fulmar (*Fulmarus glaciallis*), Arctic tern (*Sterna paradisaea*) (fig. 5.17), snow goose (*Chen caerulescens*), glaucous gull (*Larus hyperboreus*), thick-billed murre (*Uria lomvia*), and black-legged kittiwake (*Rissa tridactyla*)(fig. 7.18).

Bylot Island and surrounding waters constitute a northern oasis, particularly as summer sets in and the ice begins to break up. Parks Canada catalogs twenty-one species of mammals in the Pond Inlet–Bylot Island area, including

7.16 Odd Couple

It is not uncommon for sea gulls to cohabitate with polar bears (*Ursus maritimus*). The bears do not seem to mind, and the gulls appreciate leftovers to eat. Behind the bears is the floe edge, with fast ice (ice connected to land) in the foreground and pack ice (moving ice) behind.

caribou (*Rangifer tarandus*); five species of seals including ringed seals (*Pusa hispida*) and bearded seals (*Erignathus barbatus*); four species of whales, including beluga whale (*Delphina pterusleucas*) and narwhal (*Monodon monoceros*); walrus (*Odobenus rosmarus*); and polar bear (*Ursus maritimus*) (figs. 7.13–14, 7.16).

Without doubt, the most interesting creature found in this part of the world is the narwhal, a sea mammal with a tusk (figs. 9.04, 9,05). Martin Nweeia, a dentist as well as a medical instructor at Harvard and a research associate of the Smithsonian Institution, has been studying the unknown function of the narwhal tusk for years. He believes that the tusk, which is actually an elongated incisor tooth, is a sensory organ that may detect sound, temperature, and salinity or generate an electric current as an aid in navigation and hunting (Nweeia et al. 2007; Malkin 2005; Broad 2005). Nweeia's research on narwhals in the Bylot Island–Pond Inlet region, as well as in Greenland, involves histological studies of narwhal fossils, behavioral studies of living animals, and oral history on narwhal behavior gleaned from Inuit hunters.

Within the first decade of the twenty-first century, Parks Canada created two new parks, Torngat in Labrador and

Sirmilik on Bylot Island. Each has a large polar bear population, which inevitably creates a conflict with the human population attracted to the park. Visitors are not allowed to carry guns in Canada's national parks, although before the parks were established, visitors were encouraged—if not required—to carry a weapon. Today only aboriginal people may be armed (Kobalenko 2007), and Inuit guides may only carry a weapon to engage in traditional activities of hunting, fishing, trapping, and sealing. Could guiding activities such as hiking, skiing, climbing, kayaking be added to that list? Can guides who may not be Inuit carry guns for self-protection and fall within the definition of "traditional" use? Parks Canada regulations state unequivocally no, but people are going to get hurt and there are economic repercussions to that stance. These parks create new recreational activities with substantial direct and indirect sales and employment in Pond Inlet including airlines, hotels, restaurants, clothing, craft, and other retailers, and guides use money they make to buy and maintain snowmobiles in order to continue hunting, fishing, sealing, and trapping. The dilemma requires a more nuanced resolution.

7.17 Fair-weather Friends

King eiders (*Somateria spectabilis*) are one of many bird species that migrate between temperate and Arctic zones. They frequently travel in large flocks flying fast and low along the floe edge. Greenland Inuit used their colorful plumage and down for fancy coverlets and their down for insulation. Both became popular European trade items during the historical era.

7.18 Nesting

Black-legged kittiwakes (*Rissa tridactyla*) are numerous in Atlantic Canada and the coastal waters of Nunavut. Along with thousands of their kind, this pair nests on the vertical wall of a cliff on Bylot Island.

7.19 Snow Buntings

Snow buntings (*Plectrophenax nivalis*) live throughout the year at Arctic latitudes in both Nunavut and Greenland.

I will always treasure my memories of witnessing the summer solstice from the vantage point of the Baffin Bay floe edge. Long-tailed jaeger are common here and trailed our *komatik*, much like sea gulls trail fishermen on Maine lobster boats. Ivory gulls (*Pagophilae burnea*) and glaucous gulls (*Larus hyperboreus*) are also present at this time of year, seeking prey and fighting on a patch of ice, attacking each other with wings wildly flapping. The backdrop of the mountains on Baffin and Bylot looked as if they had been dunked in marshmallow fluff. After three days of snow, the solstice day began in near whiteout conditions, but by late evening we were basking in a golden Arctic sunlight.

Ice floating on the sea constitutes a temporary habitat for us land animals to venture upon. It is a fragile place that sparks a strange reverence in the viewer for intimacy with nature. This creation is beyond our ability as a species, with all of our technology, to control. The footprint of humanity does not dominate. Arctic ice is a place of the spirit, a place of the senses, a place of being, and a place to be with nature. It is a place that always draws me back. Here is a place where nature still controls life, but it can change, irrevocably, with just a few degrees of warming.

By AD 1500 the Little Ice Age had set in and Thule people in the central Canadian Arctic were no longer able to hunt whales in passages that were now ice-filled. These climatic changes caused Thule people to shift to hunting ring seals at breathing holes during the winter. Villages moved from the shores onto the sea ice; snow houses replaced sod houses; and winter became a time of long-distance dog-sled travel, visiting, and trade. This was the Inuit culture that the Elizabethan explorer Martin Frobisher met in 1576–78. However, along the Atlantic and Baffin Bay coasts, and in Greenland, the shift in Inuit economy and settlement was not so evident because here the ocean was open for ice-edge whaling and walrus-hunting all winter long. In these regions where the earliest European contacts occurred, the Thule whale-hunting economy continued long after European fishermen, explorers, and whalers arrived in the eighteenth and nineteenth centuries, until the stock of whales became severely depleted.

FIRST CONTACT: FROBISHER AND THE INUIT

Depictions of "skraelings" in the Norse sagas are at best ambiguous, referring to both Indians and Inuit without distinction. Real ethnography of the Inuit did not emerge until the three voyages of Martin Frobisher in 1576, 1577, and 1578 (Stefansson and McCaskill 1938; McGhee 2002). It may seem strange that Britain, with no prior history of northern exploration, began venturing into the Arctic during the Elizabethan age, but England was desperate to match (or better) Portugal and Spain's conquests in the south by finding a northwest passage to Cathay and the wealth of Kublai Khan's Orient, which had been reported a century earlier by Marco Polo. Frobisher's venture was an enormous undertaking: two ships in 1576, three ships in 1577, and fifteen in 1578—making it the largest Arctic exploration fleet of all time. Nonetheless, it failed in almost every dimension: it failed to find the Northwest Passage; its claims of discovering gold were bogus; and it bankrupted its financiers, tarnished the Crown, and landed Frobisher in debtor's prison for a spell before he was pardoned by his friend (and paramour) Queen Elizabeth. However, the expeditions achieved enormous success in geographic discovery and produced the first detailed observations of the Inuit, who—appearing Asian and wearing ornaments of copper or bronze, and using iron tools—were seen as living proof of the proximity of Cathay.

For three summers Frobisher and his captains dodged Inuit arrows—one of which found a mark in the admiral's buttocks. Undeterred, Frobisher—a giant of a man—engineered a trading encounter in which he lured an Inuit kayaker alongside his ship, *Gabriel*. When the Inuit reached for a large metal bell, Frobisher single-handedly snatched him aboard, kayak and all. On his 1577 voyage he managed to capture an Inuit man (fig. 7.20), woman, and child, whom he brought home to Bristol as living proof of the Cathay connection. The hostages all died within a month, but in the meantime the English got their first look at America's Arctic Natives when the man, wearing traditional Inuit skin clothing, demonstrated hunting ducks on the Avon River in his kayak with a bird spear. These events were immortalized in woodcuts, and in watercolor drawings by John White, that show Inuit clothing, tools, and physique in remarkable detail. White's later illustrations of Virginia Indians, painted while he was Governor of Virginia, carried the art of visual ethnography to an even higher degree.

The written descriptions of Frobisher's captains, especially those of Edward Fenton and George Best, were very informative. Best accurately described many features of the Inuit dwellings encountered in 1577, which he surmised to be winter dwellings:

7.20 Portrait of Kalicho, 1577
This illustration and a companion portrait of an Inuit woman, both painted by John White, are the earliest images of Inuit to appear in Europe. Painted from life in London before the three Inuit Frobisher captured in 1577 died of a European disease, they show the artist's attention to accurate recording of seal-skin garments, boots, and a sinew-backed bow.

Pangnirtung, "place of bull caribou," is located on the northern shore of Cumberland Sound on the east side of Baffin Island north of Frobisher Bay. As a coastal community, Pang has a place in early European history, which began with visits by the English explorers John Davis in 1585 and William Baffin—both, like Frobisher, exploring for the Northwest Passage—in the early years of the seventeenth century. In the nineteenth century it became a major British/Scottish whaling station and trading center. Compared with the length of occupation of these lands by aboriginal people, the few centuries of European presence are brief, but they were deeply consequential in cultural and ecological impacts. Whalers depleted local populations of whales and decimated caribou, walrus, and seal to feed their crews, causing hardship for Inuit residents who also suffered from introduced European diseases. Today the vestiges of the tumultuous whaling era still can be seen in Pangnirtung (fig. 7.25), which has become the center of a new industry to which local Inuit have quickly adapted: adventure tourism.

7.25 Pangnirtung Whaling Station
The Hudson's Bay Company established a whaling station at Pangnirtung in 1921, where Meeka Kilabuk's father worked when she was a child. Meeka remembers running with friends around the perimeter of the old HBC building trying not to fall on the blubber grease that was still slathered on the ground.

7.26 Arctic Idyll
Summit Lake, located at the head of the Weasel River some 40 km (25 mi) up the valley, was the destination of our hiking party. We camped here and experienced ideal Arctic weather: hours of continuous sunlight, undiminished by moist air or pollutants.

Auyuittuq National Park

By Wilfred E. Richard

HAVING REACHED NORTHERN LABRADOR, I found myself irresistibly drawn further north. I felt compelled to follow this mountainous arc of coastline and its caribou, harp seals, polar bears, as well as native cultures and European exploration history on to Baffin Island and Greenland. Pangnirtung, or "Pang," as it is called locally, is the transportation hub closest to Auyuittuq National Park, arguably the most dramatic of Baffin's four national parks. My purpose in visiting Pang in 2002 was to explore the most mountainous region of the Maritime Far Northeast.

The entrance to Auyuittuq National Park, "the land that never melts," is 25 km (16 mi) northwest of Pangnirtung Fjord. At 21,470 km² (8,290 mi²), Auyuittuq is 1.5 times the size of Connecticut. The park straddles the Cumberland Peninsula and is dominated by a north-south

7.27 Hiking in Auyuittuq National Park
Our group reached the entrance of park on Baffin Island's east coast by boat from Pangnirtung and hiked the first day over sand-covered ice punctuated by large boulders.

7.28 Stream Crossing
Hikers ferry upstream on a diagonal using trekking poles to cross a glacial stream feeding the Weasel River. Heavy packs weighing up to 35 kg (80 lbs) are carried unbuckled so if the hiker stumbles, the pack can be quickly shed.

mountain chain with Alpine-like peaks in excess of 2,000 m (6,000 ft) and has equally deep fjords. Back-country foot travel is always conditioned by weather, but here temperature and solar radiation are even more important because they determine the amount of snow melt. The greater the amount of sunlight, the greater the volume of meltwater that rushes from hanging glaciers situated in side valleys above the floor of the pass. Known as the "Alps of Canada," they draw climbers from all around the world. One of these mountains, Thor Peak, named for the Norse god

of thunder, has a vertical face of about 1,500 m (4,900 ft) and requires some of the most technical climbing in the world. This hiking route offers spectacular topographical relief with craggy granite peaks reaching higher than 2,000 m (6,600 ft.). Only a few people will ever experience these mountains, let alone know of the existence of these peaks, valleys, rivers, and glaciers, wildflowers, and birds. In retrospect, increasing public knowledge of this region was the most significant reason for visiting these sculptured mountains. It seems that in the United States—and even in Canada—geographic knowledge is on an east/west axis, not north/south. As the north is increasingly becoming the target of resource exploitation, the need for citizens of North America to gain some knowledge of the northern reaches of our continent becomes more critical.

Our trek up the Weasel River Valley to Summit Lake was 66 km (41 mi.) and had an elevation increase of about 450 m (1,500 ft.). Once inside the park, the hiker quickly reaches the Arctic Circle on a broad, till-covered plain. Here rock in its various forms and sizes dominates everything, from wind-blown sand to massive boulders and huge mountains (fig. 7.27). Campsites established beside the valley walls are always subject to chance rock falls. One particular place that hikers fear is Windy Lake, where house-sized blocks of black granite are known to cascade upon unwary visitors. Aside from a plethora of boulders lying helter-skelter, the other ever-present feature is

7.29 Weasel River Valley, Auyuittuq National Park
This view north up the Weasel River Valley captures the magnitude of the land. Tents in the lower left and a cable bridge, just barely visible where the river disappears behind an outcrop, provide a sense of scale.

7.35 Landing on Ellesmere
Ellesmere's seasonal airport at the head of Tanquary Fjord could more accurately be described as a fuel dump.

from Siberia and Alaska to Greenland. These movements coincided with warm climatic periods when the ice channels in the Canadian High Arctic islands were open and game was abundant and accessible. Trade, population movements, and communication between Greenland and Canada mostly ceased during the cooler Early Dorset period (2,500–1,500 years ago) and the Little Ice Age, ca. 1450–1750. In the near future contacts throughout the Maritime Far Northeast seem destined to blossom once again, as warmer conditions encourage the type of migrations that brought the Norse and Thule to this part of the Arctic—only now these migrations are likely to come from outsiders from the south.

Ellesmere has 50 percent of all the glacial ice in Canada, which makes it the largest concentration of glacial ice outside of Antarctica and Greenland, and its permafrost has been measured as deep as 450 m (1,476 ft). There are also snow-free zones at this latitude because for much of the year the ocean is either iced over or too cold to release the water vapor that rises, cools, and falls as snow. There is neither ice nor life in Ellesmere's desert biome (fig. 7.34), which experiences annual precipitation of less than 6.0 cm, making it drier than the Sahara Desert.

Ellesmere has been the location of many infamous Arctic explorations seeking to discover new Arctic lands or verify the existence of a theorized open Arctic Ocean. One of the largest and most disastrous expeditions was mounted by U.S. Army Lt. Adolphus Greeley as part of the International Polar Year of 1881–82. Greeley and his men were marooned at their base at Fort Conger for several years when supply ships failed to arrive, and many men died. Later, at the turn of the century, Admiral Robert E. Peary used Fort Conger as a base camp for his trek to North Pole. Peary's quest was aided by Greenland's Polar Eskimos, known as Inughuit, who acted as guides and provisioners. Peary's party harvested musk ox, caribou, and Arctic char at Lake Hazen in preparation for their trek and made a point of living off the land, Inuit-fashion, a technique that became known as "the Peary way" (Dick 2001; Fleming 2001; Malaurie 2003; Bartlett 1928).

After the Pole was "conquered," there was a surge in exploration by Europeans, Americans, and by Canadians representing the British Crown. Between the 1920s and the 1950s, and especially after World War II, the presence of Americans, Danes, and Norwegians prompted the Government of Canada to assert sovereignty over its Arctic regions. Royal Canadian Mounted Police "sovereignty patrols" cruised the Queen Elizabeth Islands, and to reinforce legal occupancy in the area the government—employing false promises and poor knowledge of the High Arctic environment—brought Inuit to settle at Grise Fjord. Today this settlement in the Queen Elizabeth Islands has a population of 150 people, but the tactics used to create the community were criticized in a Canadian Commission report that resulted in a rare official Canadian apology and reparations (Taylor 1955; Dick 2001; Waterman 2001; McGrath 2007).

Quttinirpaaq National Park

By Wilfred E. Richard

IN 2001 I JOINED TWELVE OTHER TREKKERS—Canadians, Americans, and one European—on a field trip to Quttinirpaaq National Park on Ellesmere Island. The entire island of Ellesmere measures 196,237 km² (75,767 mi²) and is more than twice as big as the State of Maine. The park itself is almost three times the size of Connecticut, making a trek here more than a simple "stroll in the park." There are no amenities or trails, only routes worked out by the trekkers themselves based on dead reckoning and planned to avoid the most severe physical barriers like glaciers, crevasses, swollen rivers, steep loose moraines, and mud. In its subarctic and Arctic manifestations, the Maritime Far Northeast is a demanding land. Reduced solar gain challenges all life forms and is disastrous to the unprepared. During the two weeks I spent on Ellesmere, other than two park personnel, a helicopter pilot, and a few geologists at Tanquary Fjord, we saw no other humans. On average, the park receives fewer than 100 people a year.

Navigation by GPS is not reliable in the High Arctic because of lack of sufficient satellites. As the plane flew north from Resolute we could see rivers of ice flowing to the sea, nunataks (mountain peaks protruding hundreds of feet through the Ellesmere ice sheet), and rock faces shaded by dull monotone grays, soft reds and browns, silently conveying the impossibility of life. We landed at Tanquary (fig. 7.35) at the southwestern entrance to the park and set out a route that would take us to within 900 km (500 miles) of the North Pole.

The High Arctic makes demands on one's body and senses that, at first, have an alien feeling. One struggles, as if in a dream, to cut through the fog of this weird, disorienting land. It was my first High Arctic experience, and I found myself ready to quit. In this clime, humility is the first lesson learned. The second lesson is that one can do more than one thinks. The third lesson is to bring less stuff. All that lightweight gear adds up: eighty pounds is eighty pounds.

We trekked about 25 km (16 mi) along the Viking Ice Cap. After descending Niagara Glacier, the weather improved as we entered the valley of the Lewis River, which flows west into Ekblaw Lake. Our spirits rose as the altitude decreased, the terrain leveled, and warmer weather and welcoming vegetation returned. As in Labrador, expanses of cotton grass (*Eriophorum*) were growing in tussocks. Rich in phosphorous and nitrogen, cotton grass is a favorite food for muskox and caribou. There were no signs of humans, but animal signs were everywhere: prints of

7.30 Glacial Retreat
At the northern terminus of our trek, we followed the Air Force River valley with its plethora of hanging glaciers. As a glacier recedes, life blooms on the minerals it has released from solid rock, as this broad-leaved willow herb (*Epilobium latifolium*) illustrates.

7.37 Hiking Hazards
This land consists of dun-colored hills of mud with the consistency of quicksand and the sticking power of pine pitch on mohair. Park authorities had recommended this detour through the valley south of Viking Ice Cap to our planned route—which, in turn, required a second detour out of the boulder clay.

7.38 Cleaves Glacier
The trekking party worked its way out of the valleys of mud and up onto the Cleaves Glacier. Not equipped for glacier climbing, hikers stepped cautiously, probing the edge of the glacier with trekking poles.

muskox (*Ovibos moschatus*) and gray wolf (*Canis lupus*), and a distant view of Peary caribou (*Rangifer tarandus pearyi*). Tufts of brown shaggy fur from muskox and white Arctic hare (*Lepus arcticus*) clung to sedges and tussocks and large amounts could be easily collected. Dorset people used to weave this fur into strong lines for boats and harpoons. We found evidence of snow geese (*Anser caerulescens*) and the long-tailed Jaeger (*Stercorarius longicaudus*). Such imprints, including our footprints, on High Arctic lands can remain for a long time.

The High Arctic is not an unending expanse of white or gray, but its diversity of plant life is considerably less than in those more southerly regions of Baffin and Auyuittuq National Park. Most common here are the purple saxifrage (*Saxifrag oppositifolia*), the yellow Arctic poppy (*Papaver radicatum*), and the white capitate lousewort (*Pedicularis*

capitata). Fields of cotton grass and Arctic poppies dominate large stretches of tundra. Buffeted by the constant winds, they behave like tiny airfield wind socks.

In the brief Arctic summer, land and sea give forth an abundance of life—sea mammals, miniature flora, and birds that migrate to these northern regions to raise their young in the highly productive but brief summer season. Ironically, in an Arctic climate, it is often easier to experience these natural wonders than in a temperate climate with its complexity of species, full vegetation cover, and human-dominated terrain.

Our route toward ice-covered Ekblaw Lake took us between two glaciers, Charybdis Glacier flowing from the north and the Scylla Glacier flowing from the south (Hattersley-Smith 1998: 30). Like the hero of Greek mythology in *The Odyssey*, we had to avoid the monsters on either side of the passageway while navigating through a morass choked by glaciated rock, ranging from rock flour to house-sized boulders. Moraines of every conceivable shape and size create a chaotic, hellish landscape, like a large sand and gravel pit peppered with obstacles hundreds of feet high. As I was climbing a moraine to photograph this field of glacial debris I fell and landed on my back with my head downhill and my pack stuck in the loose moraine gravel, as helpless as a flipped turtle.

Freed from the jaws of glaciers, we soon arrived at Ekblaw Lake, whose shore is sculpted by huge deposits of Pleistocene ice. Its lower end is blocked by Air Force Glacier, named (with less imagination than Scylla and Charybdis) to commemorate a dangerous landing made on it by a RCAF aircraft in the 1950s (Hattersley-Smith 1998: 20). At Ekblaw Lake we had reached our "furthest north." Viewing the frozen lake and its glaciers absorbed us for two days. The sun was warm, the sky a deep blue, and the glacier, pristine white.

We then turned south and followed the east side of Air Force Glacier and the Air Force River to the braided streams of the glaciated valley just north of our starting point at Tanquary Fjord. At Tanquary, I was able to board a plane and view from the air the route we had so laboriously covered by foot. Our footprints in the boulder clay were clearly visible and will probably be there for years.

Being in the silence of snow and ice in these northern lands arouses a long-sentient consciousness that was once part of our experience as a species and is still to a large degree present in peoples living along the coasts of the Maritime Far Northeast. Humans have developed a highly successful culture, but our mastery of the environment has come with a price: our disregard for the health and beauty of the land upon which we all depend for survival.

ARCTIC NATIONS CROWD AROUND THE POLAR CAP

Today air flights over the Arctic and satellite views have tamed our earlier views of the Arctic as a vast trackless expanse. This archaic view was, in part, a product of the Mercator map projection that shows the lines of longitude parallel to each other, running north and south. In fact, longitude lines on the globe converge north and south of the equator and meet at the Poles, which greatly exaggerates distances between the northern continents on our most common Mercator maps. In actuality these continents are quite close; the borders of Eurasia, the Americas, and Greenland almost meet around the Arctic. Travelers flying over the high latitudes are always impressed by how close Greenland and Iceland are to Norway and Denmark. The reality of the converging polar hemisphere forces us to reassess Arctic distances and brings a new perspective to the Norse voyages, Inuit migrations, and other Arctic explorations. As global warming opens more of the Arctic Ocean to travel, our geographic orientation to the Arctic—and to the globe itself—will undergo a profound change.

As Arctic ice melts, Canada again is showing the flag along the Northwest Passage. At Nanasivik, the site of a closed copper mine on Admiralty Inlet just south of the Northwest Passage, a Canadian naval base has been established, and there is a plan to build additional ice breakers. Issues of sovereignty are heating up again, this time fueled by global warming with its attendant reduction of sea ice, navigation access, and exploitation of natural resources. Polar nations claiming sovereignty in the north—Russia, United States, Canada, Greenland (Territory of the Danish Realm), and Norway—are becoming involved in a variety of diplomatic contests. China and Korea, unwilling to be left on the sidelines, have made bids to join the Arctic Council, the association of Arctic nations. In play is access to sea routes and natural resources that are no longer inaccessible due to ice-choked seas.

Claiming the sea floor as an extension of the topographical features of sovereign lands is becoming an international issue for all Arctic nations. Russia now is claiming the seabed from its shores to the Lomonosov Ridge in the middle of the Arctic Ocean, including the North Pole, as an extension of its continental shelf. Also being contested are the waters beyond each nation's 200 nautical mile limit, the Exclusive Economic Zone (EEZ). These limits established by the UN Law of the Sea Convention (UNCLOS) are being advanced for expansion towards the Pole by Russia, United States (Alaska), Denmark (Greenland), and Norway (Svalbard Islands). Purportedly at stake are 10 billion tons of oil and gas (Fick and Julie 2008; Wasserrab 2008). While costs and environmental hazards of exploiting Arctic oil deposits will restrict development, it seems inevitable that it will occur.

In some ways in the twenty-first century we are returning to a view of the Arctic that drove European explorers like Martin Frobisher to seek a mercantile Northwest Passage to China in the sixteenth century. Only now, that pathway leads through the middle of the polar sea, both as an Arctic transit route and for purposes of resource exploitation.

It would be wiser to seek out resources that present fewer risks, such as petroleum sites in less precarious locations and full-bore development of alternative, less centralized energy systems, predicated on locally available energy types—solar, hydro, tidal, geothermal, or wind.

7.39 Pond Inlet Cemetery
The hamlet of Pond Inlet and the mountains of Bylot Island are viewed here from the settlement's cemetery. Plastic flowers at wooden crosses meld consumer and subsistence cultures, marking graves—too many of them graves of young people who died from suicide.

8.01 Glacial Recession
The farm nestled in this green oxbow occupies land revealed when the Kiattut Sermiat Glacier receded. A millennium ago, the home of Erik the Red and his son Leif Ericsson was situated on the far side of the inlet.

8 GREENLAND–SOUTH OF THE ARCTIC CIRCLE

In the summer Eirik went to live in the land which he had discovered, and which he called Greenland, "Because," said he, "men will desire much the more to go there if the land has a good name."

(English translation by J. Sephton, from *Eiríks saga rauða*, 1880)

MOVING FROM NUNAVUT TO Greenland might at first seem like "more of the same," but in fact the two are a world apart, in geography, history, culture, and politics. The differences begin with geography. While Nunavut occupies the area between Greenland and Alaska, Greenland occupies the area between Europe and North America. Nunavut consists of many islands and peninsulas, whereas Greenland is a single landmass united by an ice cap. Great swaths of northern Nunavut are not glaciated today and consist of bare rock, whereas Greenland is a glacial landscape fringed by a mountainous coast. Culturally, Greenland has a European history extending back more than a thousand years, to Viking times, whereas Canada's Inuit contacts with Europe began with Martin Frobisher's explorations of 1576–78. Greenland has been permanently occupied by Europeans since 1721, whereas the Canadian Arctic saw no permanent white settlement until the twentieth century.

Nevertheless, there are similarities. The mountains on the east coast of Greenland near Ittoqqortoormiit (Scoresbysund) at about 70 degrees North were created by the same tectonic event that raised the Appalachian Mountains, stretching from Alabama to Newfoundland (Atlantic Geoscience Society 2001: 102). Geographers, geologists, and cartographers treat Greenland as part of North America because the island is west of the Mid-Atlantic Ridge, which passes through the middle of Iceland. Greenland and

GREENLAND

ARCTIC OCEAN

Ellesmere Island

Robeson Channel

Peary Land

Rasmussen Land

Knud

Etah
Qaanaaq

Smith Sound
Nares Strait
Kennedy Ch

Cape Sabine

Pituffik
(Thule Air Base)

Melville Bay

Baffin Bay

National Park

Greenland/ Kalaallit

Qaasuitsup

Summit Station
Location - N72°34.43', W38°28.43'
Elevation - 10,364 ft, 3,159 m

Greenland Sea

N

Constable Point
Ittoqqortoormiut
(Illoqqortoormiut)
(Scoresbysund)

Baffin Island

Upernavik

Disko Island

Uummannaq District

Disko Bay
Illulissat (Jakobshavn)
Sermeq Kujalleq

Aasiaat

Davis Strait

Cape Dyer
Sisimiut
(Holsteinsborg)
Sondrestrom Fiord

Sermersooq

Kangerlussuaq

Arctic Circle

66° 33'00" North

Iceland

Reykjavik ⊗

Kangaamiut
Maniitsoq

Qeqqata

Nuuk ⊗

Norse Western Settlements

Kulusuk
Tasillaq

Labrador

Kujalleq

Norse Eastern Settlements

Cape Farewell

Scale

0 100 200 300 miles

0 100 200 300 400 500 Kms

Nuugaatsiaq

Uummannaq District

Illorsuit

Illorsuit I.

Ukkusissat

Uummannaq Kangerlua
(Uummannaq Fjord)

Saattut

Qaarsut
Uummannaq
Qilakitsoq
Storøen I.
Uummannatsiaq
Isua
Ikerasak
Talerua I.

Nuussuaq Peninsula

Scale
0 20 miles
0 40 kms

Norse Western Settlements

Sandnes

Nuuk ⊗

Scale
0 20 miles
0 40 kms

Norse Eastern Settlements

Qassiarsuk (Brattahlid)
Narsarsuaq

Tunulliarfik (Eriks Fiord)

Igaliku (Gardar)
Narsaq

Aniaaq Fjord (Igaliku Fiord)

Qaqortukulooq (Hvalsey)

Qaqortoq

0 20 miles
0 40 kms
Scale

8.03 Flag of Greenland
Greenland's flag combines the crimson of the midnight sun, representing continuous light and radiant heat, and white representing snow and ice, the dominant features of this "green" land. The flag symbolizes a balance of light and dark, warm and cold, the forces of nature in this Arctic land.

‹‹‹ **8.02 Map of Greenland**

about two miles above sea level. The island's massive ice sheet contains 9 percent of the world's fresh water (Génsbøl 2004: 16). The weight of the ice is so great that the bedrock in the center of Greenland has been depressed below sea level, leaving Greenland's coast as a narrow band of bare rocky hills and mountainsides with scattered villages filled with houses painted in vibrant red, green, yellow, and blue.

Although Greenland holds the second largest concentration of ice on the planet, large swaths of the coast are relatively ice-free. Greenland's territory measures 2.2 million km² (836,000 mi²), but 81 percent of that land is uninhabitable ice cap, leaving only 19 percent, about 410,000 km² (158,000 mi²), for human use. This ice-free area of Greenland is about the same size as the State of California. As in many areas of the Maritime Far Northeast, the melting of glacial ice has resulted in upward "glacial rebound" of lands once covered with ice. Global warming has already begun to melt the coastal edges of the Greenland Ice Sheet. If this process continues, Greenland's coastal regions near retreating glaciers and ice sheets will be raised and relative sea level will decline, while areas on Greenland's outer coast farther away from the ice front will tip down, causing villages to sink into the sea. The harborsides in these west coast regions are already being inundated. In the long term, Greenland may come to resemble the configuration of Nunavut and Svalbard: a land of glaciated islands, bays, peninsulas, mountain ridges, valleys, and low-lying steppe-like plains.

POPULATION

Although it is the world's largest island, Greenland is only about one-third the size of Australia, which is classified as a continent. Greenland's major settlements are located on the southern part of its west coast, where the climate is moderate and the sea is mostly ice-free, the same conditions that drew the Norse to this region a millennium ago. Nuuk, Greenland's capital and the location of the Norse Western Settlement (fig. 8.04), is situated beneath the Arctic Circle and north of Cape Farewell, the southern tip of Greenland. Until 2009 Greenland was administered through a system of three counties: North (Thule), West (from Nanortalik to Upernavik), and East (all East Greenland, fig. 8.02). In 2009 these districts were consolidated into four new "kommunit" or municipalities with names designated by geographic location: Qaasuitsup Kommunia (Qaasuitsup, "the place where the light never comes," in other words, the

Nunavut have similar geological history and share many of the same species of plants and animals. Both are huge territories that extend to about 83 degrees North and have similar-sized human populations living in settlements as far north as 73 degrees. Both have been populated by members of the same prehistoric Inuit groups, beginning about 4,500 years ago. Inuit peoples continue to dominate both regions, and both have recently become invested with a large degree of political autonomy.

While most of Greenland's coast carries Arctic and Subarctic vegetation, these shores are visited seasonally by sea ice and throughout the year by icebergs. Polar currents export Arctic Ocean sea ice south along the east coast of Greenland. Because of the Coriolis effect, a force produced by the earth's rotation which causes ocean currents to turn to the right in the northern hemisphere, the East Greenland Current turns northward around Cape Farewell and flows up the west coast of Greenland, carrying both icebergs and sea ice as far north as Disko Bay. There the current is forced west by an undersea ridge and joins the ice-filled Baffin and Labrador Currents moving south along the east coast of Baffin Island, Labrador, and Newfoundland.

Greenland is dominated by a huge ice sheet (fig. 8.05) punctuated around its edges by glaciated nunataks (mountain tops protruding through an ice sheet), glacial rivers, moraines, and rock detritus created by flowing glaciers (fig. 8.01). Even when viewed from 13,500 m (45,000 ft) elevation, the horizon in all directions is nothing but ice; in fact, the horizon is often invisible. At Greenland's Summit Station located on the interior ice cap, ice elevation reaches

Arctic), with its capital, Ilulissat; Qeqqata Kommunia (Qeqqa, "in the center") with its capital, Sisimiut; Kommuneqarfik Sermersooq (Sermersooq, "the place with lots of ice") with its capital, Nuuk; and Kommune Kujalleq (Kujalleq, "the southernmost") with its capital, Qaqortoq. All of Northeastern Greenland has been designated Northeast Greenland National Park, and Pituffik (Thule Air Base), which lies within the municipality of Qaasuitsup Kommunia, is identified separately as unincorporated. What was formerly Southwest Greenland is the focus of this chapter.

Greenland in January 2010 had a population of 56,452. From 2000 to 2010, the population of Greenland has become increasingly consolidated in larger communities. These population changes are measured by growth in the administrative classifications of municipalities and towns: respectively, 11.5 percent population growth in the municipal centers (Ilulissat, Sisimiut, Nuuk, Qaqortoq) and 3.8 percent growth in the aggregate of the eighteen towns. Nuuk alone, with a population of 15,469 in 2010, now accounts for 27 percent of Greenlanders who live in Greenland. During this same time the settlements have lost 13 percent of their population.

The costs of delivering increasingly specialized systems of health, education, judicial, and other services was a major factor stimulating the move toward consolidation of services. In recent years housing booms occurred in the capital, Nuuk, and the regional center of Ilulissat, and the island-wide "one price" system was discontinued, at least in part. With the cost of delivering goods factored in, living expenses in the smaller settlements have risen dramatically. An analogy would be if standard first class postage rates in the United States were replaced by a distance scale, that is, the greater the distance, the greater the price of a stamp. Similar centralization or consolidation strategies have been used in Nunavut, Newfoundland, and Labrador.

8.05 Greenland Ice Sheet
Greenland has the largest ice sheet in the Northern Hemisphere. It is shaped like a large saucer with coastal mountains that restrain a massive pool of ice nearly two miles thick. The glaciers exit to the sea in a few places via deeply cut fjords.

8.06 Kunuunnguaq Fleischer
The project director for Greenland's Department of Education and Research explained the peculiarities of the Greenlandic language.

Our focus in this chapter and the next is largely on the municipalities or towns, villages, and smaller settlements arrayed along Greenland's western coast, where the majority of Greenland's population lives. Unless otherwise stated, population data are from Thomas Brinkhoff (2009) and Statistics Greenland (2010). When the term "district" is used, it applies to a town and its neighboring settlements. For example, the town of Uummannaq includes its satellite settlements of Qaarsut, Saattut, Ikerasak, Nuugaatsiaq, Illorsuit, and Ukkusissat.

LANGUAGE

While the Inuit of northern Canada have adopted one word, Nunavut, "our land," to identify their homeland and the Inuit of Labrador use Nunatsiavut, "our beautiful land," in Greenlandic, morphology and sounds are not so simple. The official language of Greenland, Kalaallisut, is a northwest Greenland dialect, which is spoken by about 93 percent of Greenlanders. Another dialect, Tunumiisut, is spoken in East Greenland by about 6 percent of Greenlanders (Tersis and Taverniers 2010). Kalaallit Nunaat, "Land of the Greenlanders," combines Nunaat, meaning "land" in Kalaallisut, and "Kalaallit," the term by which Greenlanders have been known since the early 1900s.

Orthography, that is, writing words with the proper letters according to standard usage, or more simply, the representation of the sounds of a language by written or printed symbols, is also complicated in Greenland. Many variations of proper Greenlandic names were given when I asked for spelling of Greenlandic words in the field. Even on maps (including Google) and charts, there is little agreement and orthographic sloppiness is rampant. Uncertain of the correct spelling in making maps of Greenland, I turned to educator Kunuunnguaq Fleischer (fig. 8.06) for authority, which is where I should have started. He explained that Samuel Kleinschmidt, whose spelling from 1851 presupposed a thorough knowledge of the grammar, used different ways to clarify long (^) and short (´) vocal sounds. He used a tilde (~) when a long vocal sound met a double consonant, but today that convention has been eliminated, so Kleinschmidt's *āmais* is written *aamma*, and *sātut* is now *saattut*. Accented vowels have been replaced by aa, ii, ee, oo, uu, and K is usually replaced by q. Kleinschmidt used K, which in different European texts was written as a k or K or K', so in many maps and books the spelling was very random and incorrect. In some cases q was used combined with Kleinschmidt's spelling, even though q was not introduced until 1973. So, for example, Umánatsiaq (now correctly written Uummannatsiaq) is incorrect in both Kleinschmidt's spelling and the spelling of 1973. Most Danish typewriters simply did not have the type needed for the correct spelling, or the Danish/English writers did not have the knowledge to write the correct names.

I have tried to be accurate according to modern usage, and to facilitate understanding by outsiders, most non-Greenlandic (usually Danish) place names are preceded by their original Greenlandic names: Kangerlussuaq (Søndre Strømfjord), Nuuk (Godthåb), and Ilulissat (Jakobshavn). The suffixes *-miut* and *–vik* occur frequently in place names throughout the Arctic. Where varieties of Inuktitut are spoken, the suffix *-miut* means "the people of," as in Nunamiut, "the people of the land," from *nuna* (land) and *miut* (people of). *Vik* connotes place or land, as in Nunavik, "the land place." Kunuunnguaq explains, "For me Nunavik means the Real Land, but *–vik* is also used in words like *ilisivik*, "the place where you place something"; *aasivik*, "the place where you spend the summers"; or Upernavik, "the spring place." But it can also mean "authentic, real" as in *angutivik*, "a real man"; *kuultivik*, "real gold"; or *pitsavik*, "really good." In Nordic languages *-vik* has the same meaning as *nuuk* in Greenlandic. John Houston (1995: 60) writes humorously about *miut* in giving an Inuit perspective to Western concepts, coining the words "missionarymiut" and "governmentmiut."

Greenlanders are ambivalent about using the term "Inuit." They do not explain their nationality as Eskimo, but rather as Kalaallit, "people of Greenland." Yet, as Greenlanders, they possess a feeling of solidarity with other Inuit across Canada and Alaska who use the term Inuk (singular) or Inuit (plural), which originally (that is, in their own languages) meant "human being." In short, the Inuit people of Greenland make a distinction whenever possible between the classifications of Inuit, Eskimo, and Greenlander, and prefer to be called Greenlanders in their daily lives. Most Greenlanders consider the term "Eskimo" as somewhat derogatory, so they prefer "Inuit" when referring to the broadest distribution of their people.

Based on her research in the Canadian High Arctic village of Pond Inlet, Katharine Scherman (1956) evaluated the integration process between European and Inuit cultures. She believed that in the early 1950s Danes and Inuit/Greenlanders had achieved a more successful cultural blend than in Canada or Alaska. Such comparisons are very difficult to make because of different histories and political arrangements. Alaska has a contentious ethnic milieu, with Euro-Americans frequently pitted against the state's Alaska Native population over subsistence rights, mineral exploitation, wildlife, and other issues. Canada has a relatively short history of Euro-Canadian governance, with Inuit having only recently obtained basic governance rights. According to the 2011 Census of Canada, 83.6 percent (24,640) of the Nunavut population is Inuit. Statistically anyone born in Greenland is Greenlandic, but, unofficially, there are distinctions. In 2000 there were 6,755 Danes in Greenland, a figure that represents about 12 percent of population, leaving a Greenlandic population of 88 percent. In 2007 almost 13,500 Greenlanders lived in Denmark. As one Greenlandic friend commented, Greenland remains the only polar-centric political entity in which the indigenous population is in the majority, as compared with, for example, Canada or Alaska, where Inuit or Native Americans are in the minority.

The social acceptability of Danish–Greenlandic intermarriage and of Greenlandic students in Denmark pursuing education beyond high school are both factors in this successful integration and ethnic/racial evolution of the local population. In 2010 about 70 percent of Nunavut's population is Inuit, compared with about 88 percent of Greenland. The distinction between indigenous and introduced populations seems less divisive in Greenland than in Nunavut. Greenlandic culture has become an amalgam of traditional Inuit interdependence with the land and Scandinavian communalism.

ECONOMY

Greenland is a self-governing territory within the Danish realm, or Rigsfællesskabet, which consists of Denmark, the Faroe Islands, and Greenland. In this respect, its relationship to the Danish realm is similar to Nunavut's relationship with Canada. What is not similar is that Greenland continues to move toward independence as a sovereign state. On National Day, June 21, 2009, thirty years of home-rule ushered in a new era of self-governance, and the Home Rule Government of Greenland became the Government of Greenland. One implication of this shift toward independence is that Greenland's economy must replace the annual subsidy from Denmark, DKK 3.2 billion (US $640 million) in 2013, equivalent to $11,200 per capita. Greenland's GDP in 2011 was US $2.3 million, of which most is derived from fisheries. If all subsidies are terminated, Greenland will need to find most of the shortfall from other sources than fisheries. In the future, if fisheries decline from global warming as much as projected, proceeds from natural resources—petroleum, minerals—and tourism will become much more significant factors in the Greenland economy.

In the warm period during the 1930s a thriving cod fishery extended far north into Davis Strait, but cooler waters and overfishing in the North Atlantic resulted in its collapse. Cod are now being caught in small amounts as far north as Ilulissat, and perhaps with climate warming and careful regulation it may again become an important resource. Like the fisheries in Nunavut and the Maritimes, some fish stocks are stressed in specific locales. Sealing used to be the basic sustenance of most village populations, especially those in northern parts of Greenland. Today with Euro-American embargoes on seal products, fishing is replacing sealing as the dominant extractive activity in the settlements.

To pay for the costs of independence Greenlanders have been forced to initiate natural resource development, including aluminum smelters, oil exploration in Disko Bay and other offshore locations, and mining of gold, diamonds, and other minerals on the Nuussuaq Peninsula and elsewhere. The discovery of favorable geological formations and a reduction in the severity of ice in northwestern Baffin Bay has stimulated oil and gas exploration. Projected offshore reserves are estimated at 1 percent of world's proven oil reserves and 2 percent of all proven gas reserves. But, at this time, proven reserves of a marketable size have yet to be found.

In Ivittuut, south of Nuuk, cryolite, which was used as a catalytic agent to produce aluminum, became a strategic Allied resource that was protected from the Germans during World War II. That mine has since played out, but in the near future an Alcoa aluminum smelter may be built at Maniitsoq, between the towns of Nuuk and Sisimiut, powered by glacial meltwater (Knudsen and Andreasen May 2009). Another mine, Maarmorilik (Black Angel), located in Uummannaq Fjord, was closed in the 1990s. However retreating ice has uncovered a new vein containing lead, silver, and zinc, so Black Angel may open again. Nalunaq, located just to the west of Cape Farewell, is an operating gold mine. In the hills above Narsaq, an outcrop of rare earth metals and uranium has been recently discovered. Greenland's Nuna Minerals has affiliated with the global enterprise Rio Tinto Mining which is focused on Greenland's gold, diamonds, rare earths, and iron. At present Greenland's "zero-tolerance" policy on uranium mining has restricted uranium mining in Narsaq to a continuing debate, which has yet to be resolved. While the Government of Greenland approved exploration in late 2010, as of mid-2012 actual mining has been held up by its "ban on mining radioactive elements" (Weaver 2012). At a smaller scale, materials like tiny rubies are now being mined and incorporated into jewelry carved from bone and antler for the tourist trade.

Tourism is one of the growth industries that can provide economic stability in the settlements, but its development is hampered by limited investment opportunities and the need for most support amenities to be imported. There is a place for the amenity-based tourist in Nuuk, Sisimiut, Ilulissat, and Qaqortoq, and the more adventurous traveler, who seeks a back-country, cultural, or recreational experience, will find much of interest in the small rural settlements. In 2008, 237,000 nights were spent at hotels, with almost one-half of the rooms taken by Greenlanders and one-third by Danes; non-Danish Europeans and Americans each contributed about 3 percent. This split by nationality is characteristic of retail commercial activity in Greenland.

GREENLAND AND THE UNITED STATES

Connections between Greenland and the United States have been long and frequent. In the early nineteenth century, whalers from New England worked the waters of Baffin Bay, while American explorers like Robert Peary, Elisha Kent Kane, and Charles Francis Hall ventured north to map Greenland's polar regions. In 1867, after the US Civil War, Secretary of State William Henry Seward attempted to expand the northern territory of the United States by purchasing lands that held promise as strategic regions or for their fisheries or other natural resources. Although he succeeded in purchasing Alaska from Russia, Seward lacked sufficient support and confidence from the US Congress to acquire Greenland from Denmark.

In the early twentieth century, following his Greenland and North Pole explorations, Admiral Robert E. Peary petitioned the US government to claim sovereignty over northern Greenland. Like Canadian endeavors to ensure sovereignty of its Arctic territory by establishing a colony of Inuit at Grise Fjord in the High Arctic, the Danish government in 1925 had moved a group of Inuit to distant Scoresbysund "in order to counter a Norwegian claim to portions of the East Greenland coast" (McGhee 2008: 109). In 1933, Knud Rasmussen, who had become famous in Denmark as an explorer and popularizer of Greenland culture—by leading the famous Thule Expeditions, which initially collected and published Inuit stories and oral literature and later expanded into broader geographic and scientific research, and for his historic dogsled trip from Baffin Island to Alaska—supported

8.07 Cooperation during World War II
The airbase built in south Greenland in the area of the eastern Norse settlements served the Great Circle Route and was known in military code as "Bluie West One." Now it is Narsarsuaq airport, linking Greenland with Iceland and Denmark.

NARSARSSUAK AIR BASE GREENLAND (BW-1)

Dedicated to a joint effort by the United States of America and Denmark
July 1941 - August 1958

11 November 1996

8.08 National Science Foundation Transport
An LC130 cargo plane is modified by skis that surround the regular wheels, which enables this heavy-lift craft to land on snow.

8.09 Aboard NSF Aircraft to Greenland
En route to Kangerlussuaq airport, which serves as the staging, supply, and administrative center in Greenland, scientists read and rest. It is serviced by the 109th Air National Guard flights chartered by the National Science Foundation.

8.10 Summit Station
One of the research sites operated by the National Science Foundation is at Greenland's highest point, Summit Station, elevation 3,159 m (10,364 ft), nearly 3 km (2 mi) above sea level.

the Danish Crown in its effort to expand its sovereignty to include Eastern Greenland. The Danes were generally less opposed to the American proposal than to permitting Norway's hegemony in Greenland. But Rasmussen, the "father of Greenland," was repulsed by the thought of a divided Greenland. "By appealing to his Native background and his extensive knowledge of the diverse communities of Greenland as a whole, he established Danish sovereignty over the entire island. He, Knud [Rasmussen], just knew that they were one people indivisible" (Hastrup 2007). The International Court of The Hague ruled that all of Greenland should be a territory of Denmark, expunging all legal claims by Norway to the former Norwegian colonies in Eastern Greenland.

World War II and the German occupation of Denmark caused the United States to establish a non-Axis government in Greenland. In cooperation with Greenland's Danish Governor Eske Bruun and the Danish Ambassador to the United States Henrik Kaufmann, air bases were built to facilitate the transfer of aircraft and goods to the European theater. After the war, Thule Air Base was constructed in northwest Greenland to serve as a forward base countering the Soviet threat. Part of this arrangement involved moving the Inughuit (Polar Inuit) population from their settlement at Thule north to Qaanaaq. In the 1990s, after the end of the Cold War, all US bases except Thule, which is still controlled by the US, were returned to Denmark. The issue of moving the Inughuit settlement from Thule remains a thorn in Greenland's relationship with Denmark and the United States.

Today, the primary links between Greenland and the United States are military and scientific. The military focus is dominated by the operation and maintenance of the Thule Air Base and logistical support for scientific research on climate change and other projects managed by Greenland Kangerlussuaq International Science Support and the US National Science Foundation (NSF) (figs. 8.08–8.10). One of NSF's current research projects in Greenland is maintaining a "Flux Station" at Summit Camp, located in a lab carved beneath the snow pack. This project measures the chemical composition of airborne organic chemical compounds in the snow mass. Like albedo measurements of ultraviolet light, gases are both absorbed and reflected by snow. The flux station measures the chemistry of gases that are absorbed and reflected between the atmosphere and the snow pack (Lars Laurens-Ganzeveld, pers. comm., 2009). They have learned that natural and anthropogenic emissions are reaching the Greenland Ice Sheet from "fossil fuel combustion and biomass burning [which includes forest fires and farmers burning off fields], both in ambient sampling and as a paleo-record of previous source activity" (Hagler et al. 2008: 2486).

PREHISTORY AND NORSE GREENLAND

Greenland's pre-European history has been researched by archaeologists, of whom the first was Ole Solberg (1907) who (mistakenly) saw similarities in stone tools between West Greenland's earliest culture, Saqqaq, and European Paleolithic cultures. Nuuk and its surroundings, like Southwest Greenland, were

8.11 Tjodhlide's Chapel
Erik the Red's wife built a diminutive Christian church in Brattahlid, Greenland. But Erik refused to become a Christian.

8.12 View from Hvalsey church
The church was built 900 years ago with meter-thick stone walls. Its doorway frames some of the outbuildings at Hvalsey, including residential quarters, cow barns, and a feasting hall.

8.13 Norse House
This reconstructed residence features sod walls with upright log trusses, central fireplaces, and lateral sleeping platforms. Similar longhouse architecture helped identify the three houses at the L'Anse aux Meadows site in Newfoundland as Norse.

settled first by the Paleo-Eskimo Saqqaq culture about 4,000 years ago, followed by Dorset people around 2,500 years ago (Fitzhugh 1984; Gulløv 2004). When the Norse arrived in southwest Greenland about AD 1000 they found the land abandoned but noted signs of former inhabitants. Toward the end of the Norse settlement period, which in the Western Settlement was about 1350, Thule culture began to arrive from the north and interacted to some degree with the Norse. Although there has been speculation about warfare, trade, and genetic mixing between Inuit and Norse, relatively little contact occurred before the Norse abandoned the region as the Little Ice Age climate advanced and livestock farming became impossible. After their departure, the Inuit mined Norse sites for metal and other useful materials, and a new culture known as Inussuk appeared, which was essentially a Thule culture to which some Norse raw materials were added. The descendants of the Inussuk people are the Greenlanders living in the Nuuk region today.

European history in Greenland began in the waning years of the first millennium when the Norse Eastern and Western Settlements were established by Erik the Red in 985. Østerbygd, the Eastern Settlement just northwest of Cape Farewell, had three times the population of Vesterbygd, the Western Settlement, which was located in the region around Nuuk, about 300 km (180 mi) to the north. These two aggregations included a total of 278 farms, seventeen churches, and two monasteries (Ingstad 1966: 24; Fitzhugh and Ward 2000). Both regions share a complex geography of peninsulas, fjords, islands, and natural harbors. Its most sheltered locations deep in the fjords provided relatively mild temperatures and more vegetation for cattle and sheep than found along Greenland's outer coast.

The first Norse settlements were little more than groups of farms occupying the most favorable inner-fjord agricultural regions. The Norse did not establish villages in the traditional European sense. Rather, their farms were dispersed kilometers apart from each other along the shores of the fjords near a chief's farm that (eventually) included a church. Even the few churches that were stone lacked the monumental scale of European architecture, but they still convey a strong sense of European culture not present in the raw, biting Arctic lands that constitute much of Greenland and Nunavut. In fact, Southwest and Western Greenland are climatically exceptional for the high northern latitude. The distance from these Norse-founded settlements to the northern coast of

8.14 Hiking Kiattut Sermiat Glacier
The remnant of the glacier that has receded from Narsarsuaq and the surrounding hills has become a popular trekking area.

Greenland is about 2,600 km (1,600 mi)—all of it due north. This distance is the primary reason for the dramatic climatic differences from one end of Greenland to the other.

Southern Greenland is about 900 km (560 mi) east of Labrador at 60 degrees North. In warm climatic periods, warm air associated with the Gulf Stream (also known as the North Atlantic Drift) make livestock raising and limited agriculture possible in the protected fjords of southwest Greenland, whereas comparable latitudes in Labrador remain under the icy grip of the Labrador Current. Norse settlement began during a warm period known to climatologists as the Medieval Warm Period. The Norse found the inner reaches of the fjords green, unoccupied, and suitable for their traditional economy of raising sheep, cattle, pigs, and horses. Norse had occupied all suitable farmlands in Scandinavia and Iceland, and population pressure and unrest was mounting. By trumpeting the empty land's greenness and suitability for grazing, Erik convinced many impoverished farmers from Iceland and other northern European Norse lands to follow him to Greenland.

The first Norse farms were established in the Eastern Settlement and are now known as Narsarsuaq, Qaqortoq, Qaqortukulooq (Hvalsey), Upernaviarsuk, Narsaq, and Qassiarsuk. In Norse times the latter was known as Brattahlid or "Erik's Fjord." Soon after, farms were begun in the Western Settlement around the present Greenland capital, Nuuk. All of the Eastern Settlement Norse farms are contained today within Kommune Kujalleq. Nuuk and other Western Settlements are now within Kommuneqarfik Sermersooq. Although these are only a few of the settlements originally established by the Norse and documented by archaeological research, they are among the most important sites of Greenlandic Norse culture.

Humanity is experiencing a dramatic change in its history as a result of the industrial age and the shift from being passive witnesses to agents of global change. In the 12,000 years since the end of the last ice age, a geological period known as the Holocene, we fossil-fuel addicts have given birth to the Anthropocene (age of man): we have supplanted the normal geological and astronomical agents governing humanity's previous history. Our agroindustrial world is now solidly based on the burning of hydrocarbons, and these gases are changing the nature of our world by elevating the concentration of CO_2 and other warming gases that cause glaciers to melt, sea level to rise, and oceans to acidify. Far from being immune

191

to these changes, the Maritime Far Northeast will soon find itself a portal—possibly a gatekeeper—to a new previously closed sector of the planet. Besides being on the front line of climate change, it is soon to experience increased mining and industrial activity with the opening of once-closed Arctic Ocean sea routes between Asia and the North Atlantic. No longer isolated and cut off by sea ice from the outside world like the fourteenth-century Greenland Norse, the MFNE will experience a dramatic increase in exploitive industry and trade, becoming a portal to a new more interconnected world as part of a new geological epoch, the Anthropocene.

Eight trips I have made to West Greenland from 2000 to 2012 structure this narrative, which provides social and ethnographic records and information on climate and geographic conditions in the first decade of the twenty-first century. Documenting these northern regions and promoting knowledge about them to the wider world has become more important today than ever before.

NARSARSUAQ

The international airport at Narsarsuaq ("big plain") is located immediately across Tunulliarfik (Erik's Fjord) from Qassiarsuk (Brattahlid), Erik the Red's home settlement at 61° 11'N, 45° 22'W. Cape Farewell, the southern tip of Greenland, lies to the southeast. In 1941 the United States constructed an airfield on the great circle route to Europe to aid the war effort, and to many it became known by its military code name, Bluie West One (fig. 8.07). Other bases in this chain—Gander, Newfoundland; Goose Bay, Labrador; Iqaluit, Nunavut; Keflavik, Iceland—are now the stepping stones that unite peoples and cultures across the vastness of the Maritime Far Northeast.

While at Narsarsuaq I trekked into the hills along the Kiattut Sermiat Glacier and made a steep ascent

8.17 Fields and Icebergs, Narsaq
To the southerner the juxtaposition of icebergs with potato and hay meadows is bizarre, but not to a Greenlander, or a Viking.

8.18 Igaliku
Gardar was the largest of the Norse settlements and, as the bishop's seat, the most important, but it too had been abandoned by the mid-fifteenth century. Beginning in the eighteenth century, Greenland was resettled, and Gardar became Igaliku, a modern farming community that also caters to travelers.

using fixed ropes to a spot overlooking the interior ice sheet. This ice is the toe of a glacier, which at some ancient time had flowed past Narsarsuaq into Tunulliarfik (Erik's Fjord). From here I could see an agricultural feature of the glacial valley below—a farm under tillage on the fjord's left bank. Farming is not limited to ice-free fjords. Flying over the Eastern Settlement area, one sees scattered pastures and hay-fields on hillsides and even on hilltops.

QAQORTOQ

The port town of Qaqortoq, pronounced, kra-kror-tok, meaning the "white place," is the capital of Kommune Kujalleq, and located about 120 km (72 mi) by water from Narsarsuaq. We cruised down Erik's Fjord, navigating around icebergs, rocks, and islands, and enjoying an excellent view of the mountains. We landed in Qaqortoq, the commercial center of Southwest Greenland, which has a population of 3,526 in the town and adjoining settlements. The settlement fills the valley and flows up toward the surrounding ridges with red, blue, green, and yellow frame houses, set off by orange and yellow arctic poppies, and inhabited by friendly people. The image reminded me of the vivid colors of Scandinavian villages and the brilliant hues of homes on Jelly Bean Hill in St. John's, Newfoundland. Northern residents, it seems, love color!

Qaqortup Katersugaasivia, the local museum, has an excellent bookstore, an exquisite collection of kayaks, photographs, and women's traditional dress, as well as many *tupilak* carvings, usually from sperm whale teeth, which are representations of grotesque evil spirits. *Tupilaks* used to be connected with sha-manistic beliefs but now have become popular items for the tourist trade. The history of the museum was also impressive: the "blue room" was used by Charles Lindbergh in 1933 when he was surveying Greenland's flight conditions for Pan American Airways, and the "red room" on the third floor, at the top of very steep, narrow steps, was used by Knud Rasmussen from

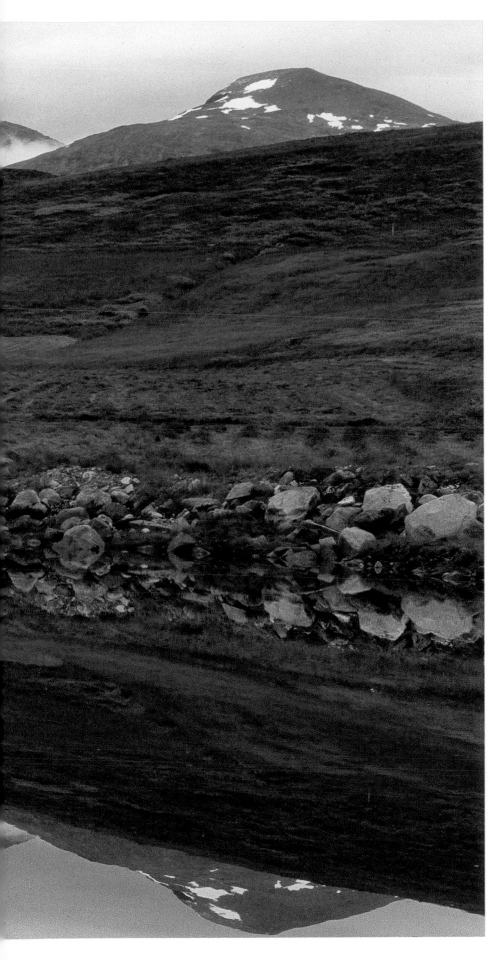

1930 until he died in 1933 from eating tainted meat on a trip to East Greenland.

A steep winding roadway lined with brightly colored homes and flowers leads to the town wharf, where we boarded the *J. F. Johnstrup*, an old, well-appointed wooden boat of 15 m (50 ft), bound for Hvalsey.

QAQORTUKULOOQ (HVALSEY) AND UPERNAVIARSSUK

We cruised northeast 20 km (12 mi) between the island of Arpatsivik and the Upernaviarsuk Agricultural Research Station to Qaqortukulooq (Hvalsey) Fjord and the Qaqortukulooq Church (figs. 8.12, 8.15). The church, built in the early fourteenth century, is the best preserved Norse structure in Greenland. Measuring about 16 m (52 ft) long, its rock walls are 1.5 m (4 ft) thick, but its roof, probably originally of wood and sod, has disappeared with the ravages of time. Adjacent to the church is a large banquet hall, still standing, and the foundation of a multi-roomed house. Barns for cattle, sheep, and goat with adjacent hay storage, storehouses, a horse pen, and a well complete the trappings of a rich farmstead. Saga accounts suggest its first owner was Thorkell Farsek, Erik the Red's cousin.

Today all except the stone church and banquet hall are a jumble of rocks and sod. A wedding between Sigrid Bjørnsdottir and Thorstein Olafsson that took place in Hvalsey church in September, 1408, was the last event to be recorded in Greenland Norse history, eighty-four years before Columbus reached the Americas.

During their tenure of about 400 years, Norse agricultural practices seriously degraded Greenland pastures. Within the first few decades Norse settlers had harvested the small patches of wood and brush for fires and charcoal, and during succeeding generations pastures were overgrazed and the land stripped of sod for building material. The beginning of the Little Ice Age in the fourteenth century coincided with the maximum size of the Norse population and its livestock holdings (Lynnerup 2000; McGovern 2000; Diamond 2005). As the climate worsened, the Norse diet shifted toward greater use of marine-based resources, especially fish, seals, and walrus.

8.19 Itelleq, "The Crossing Place"
The juxtaposition of glaciers and raw, ancient rock with modern agricultural practices is a reminder that although only a small sliver of this land can be farmed, it has been tilled for more than one thousand years.

Isotope studies of Norse burials have shown that in the early settlement period 80 percent of Norse nourishment came from the land and 20 percent from the sea. Three centuries later, the percentages were reversed, causing researchers to reconsider the theory that the extinction of the Greenland Norse resulted from their failure to adopt Inuit hunting and foraging techniques. But even as they made better use of marine resources, the Christian Norse were not about to adopt the life of the "heathen" Inuit, becoming nomadic hunters and fishermen when all of their cultural and religious values were doggedly Christian and European. Culture as well as climate may have doomed the Greenland enterprise.

To avoid the Norse excesses of the former Eastern Settlements, an Agricultural Research Station has been established in Upernaviarsuk to ensure that sustainable agricultural science would be practiced from now on. Like the Norse, Greenland farmers today grow hay and graze more than 20,000 sheep and a few horses and cows, as well as 2,500 domesticated reindeer. They also cultivate cucumbers, beans, flowers, carrots, tomatoes, peas, and potatoes, both outdoors and in greenhouses. However, there is a difference: these farmers are Greenlandic, not Norse or European; they have succeeded in doing what the earlier Norse could not do—adapt to a vastly different economy and culture. Warming climate is already having a positive impact on this rapidly developing agricultural sector of the Greenland economy (Jenkins 2010). But even a warm climate, modern technology and greenhouses do not make cultivation easy or predictable in Greenland.

Vegetables and many other food products are distributed in Greenland by the Pilersuisoq stores, a chain found at some sixty locations throughout Greenland and advertised by a smiling polar bear logo. It is also known as KNI (Kalaallit Niuerfiat), formerly the Royal Greenland Trading Company or KGH. In addition to being the national supplier and retailer of fuel products, Pilersuisoq distributes food and has exclusive rights for beer and soft drinks. In every situation where we prepared our own food, Pilersuisoq was our source of "daily bread," a repast consisting of crackers, sweets, goat cheese, apples, Chilean wine, and occasionally an expensive treat like a Greenlandic cucumber grown at the Agricultural Research Station.

IGALIKU (GARDAR)

Gardar, Old Norse for "farm," is today known as Igaliku, "deserted cooking site." Toward the end of the last millennium, the periodic assembly of Norse at which the customs and practices were debated and

8.21 Arctic Chromatic
Kangaamiut, with its eye-stopping, vividly colored housing, displays a palette of primary colors that brightens the landscape of endless gray rock.

8.22 Mother and Child, Kangaamiut
When a tour group arrived in Kangaamiut, a mother and child greeted us as Canadians because the ship was flying the maple leaf.

enshrined as law—the Althing, which some have called a proto-parliament—was held at Gardar. In 1124 Gardar became the seat of the Christian Bishop of Greenland and has become a place of historical pilgrimage much like Thingvellir, the famous Norse assembly place in Iceland.

Itilleq, "the crossing place," is a sheep farm (fig. 8.19) that, from a distance, with its bales of hay wrapped in bright white plastic sheeting, could have been a farm in Maine, Quebec, or Maritime Canada—except for the snow-capped mountains and bergs in the fjord. From Itilleq a 3 km (1.8 mi) gravel road ascends a hill, then drops to the other side of the peninsula at Igaliku, where the ruins of the church of the Norse Bishop of Greenland are found. Seeing a farmer with sheep and dogs and children riding Icelandic ponies across his pasture reminded me of an incident reported in the Norse sagas that occurred across the fjord in Qassiarsuk (Brattahlid). One thousand years ago, as Erik the Red and his son Leif were preparing to embark on a voyage to Vinland, Erik's horse stumbled on the way from his house to the boat. In Norse belief, a horse's stumble was a bad omen, and Erik was enough of a believer to remain home, thus losing the chance to discover Vinland and America.

9.01 January Light over Uummannaq Fjord
During the cold season, Greenlanders forge the bonds of community with enthusiastic socializing. A lack of land-fast ice in recent years has often made it impossible to cross Uummannaq Fjord by sledge, the traditional way.

9 GREENLAND–NORTH OF THE ARCTIC CIRCLE

Most people…might be oppressed by such surroundings, with its silence and inhuman expanses…But he who seeks peace and quiet in Nature, undisturbed by human activity… will find here [in Greenland] what he seeks…even although, beset in the ice, one is a plaything of the forces of Nature…it is the wild and torn side of Nature that exerts the greatest power over the senses

Roland Huntford, *The Explorer as Hero*

WHEN ONE HEARS OF THE ARCTIC, the mind's eye rushes to a vision of an ice-bound panorama with polar bears and walruses, of travel by dog sledge, and of Inuit living in igloos, turf houses, or tents. North of the Arctic Circle this vision—at least historically—is somewhat congruent with reality.

This is the world that Knud Rasmussen, Arctic adventurer, filmmaker, and writer, who was widely acknowledged as the spiritual and literary father of Greenland, explored from the turn of the last century. In 1910 Rasmussen and his friend and expedition partner Peter Freuchen, who achieved fame for his scientific and popular writings about the north, established a trading post at Cape York in Polar Eskimo territory at a site they called Thule. Freuchen coined the name, adapting a term from the Classical and medieval literature, "Ultima Thule," which originated with Pytheas, a Greek explorer who was describing the northernmost regions of the known world. Freuchen chose the name (1953: 67) because the location was "north of everything and everybody." With profits from the trading post, Rasmussen and Freuchen launched a series of "Thule Expeditions" with the initial purpose of collecting and publishing Inuit stories and oral literature. Later the expeditions expanded into geographic and scientific areas. The Fifth Thule Expedition (1921–25) became famous for its anthropological studies and for Rasmussen's dogsled trip with two Inughuit companions across northern Canada to the Bering Strait. As a result of these explorations and the voluminous scientific publications he and his colleagues produced,

Crafts, lodging, restaurants, and transportation are all part of the highly differentiated, labor-intensive tourism sector are now spreading in Greenland. Increased focus on tourism throughout the Maritime Far Northeast has brought new products and services to the market. Thirsty visitors attracted to the North to experience traditional cultures, Arctic animals, and the effects of global warming and melting ice may now purchase products like Greenland Brewhouse Pale Ale and Siku Glacier Ice Vodka, as well as bottled glacier water. Greenland's greatest resource is its proven reserves of water (stored as ice), which will be in demand by an increasingly thirsty planet and as a source of hydroelectric power needed for a growing population and proposed aluminum smelters and other industrial developments.

SISIMIUT

Sisimiut, the capital of Qeqqata ("in the center") Commune, is at the southern boundary of West Greenland and sits on the Arctic Circle. Ecologically,

West Greenland is identified as Middle Arctic, a region that extends north from the Arctic Circle to Etah, a community in far northwestern Greenland, (Mortensen 2001: 239). This huge area has been divided for administrative purposes into West and North Greenland, with Qeqqata and Qaasuitsup Communes spread along the west coast of Greenland across six degrees of latitude from Sisimiut (N66° 56'; W53° 40') to Upernavik (N72° 47'; W56° 09'), about 670 km (400 mi) or one-quarter of the length of Greenland. A majority of the island's population resides here.

Sisimiut, formerly Holsteinsborg, is important in the human geography of Greenland for two critical reasons. First, it is the southernmost location in West Greenland where the husky dog is found. Second, according to local sources, water temperature and currents have made Sisimiut the southern boundary of permanent sea ice. Dogs are still required for hunting and transport north of Sisimiut, where ice is more reliable, which also makes Sisimiut the de facto southern boundary of traditional Inuit

9.07 Icebergs in Uummannaq Fjord
Light plays on icebergs, creating a a glittering panorama of sky, sea, and ice. Deep blue indicates greater ice compression than white, which contains large bubbles of uncompressed air when it fell as snow. Blue ice is super-compressed snow ice or a glacial stream that froze.

9.08 Craftsman Johannes Mønch
Johannes Mønch, an artist from the High Arctic, possesses some of the striking physical features, including high cheekbones, of northern Greenland people.

culture in western Greenland. Maniitsoq, just south of Sisimiut, is the northernmost point to which dogs other than huskies can be imported, a rule established to maintain a viable husky breed in the northern settlements.

The region from Sisimiut to the Nuussuaq Peninsula on the north side of Disko Bay was called the Nordsetur, "northern dwelling place," by the Greenland Norse, although that term is somewhat of a misnomer as this area was not permanently settled, but only visited seasonally by the Norse. This coastal stretch was the southernmost hunting grounds where walrus, narwhal, and polar bear could be caught, supplying Norse needs for ivory, meat, fat, and hides, both for local consumption and export to Iceland and Europe. Much of the ivory was dedicated to meeting the tithing requirements of the Church in Greenland and Norway. These hunting grounds are probably where Norsemen first met Thule Inuit, who began migrating south from Northwest Greenland into West Greenland in the 1200s–1300s.

9.17 Uummannaq

The mountain known as Uummannaq, Greenlandic for "heart-shaped," provides the signature landscape of this island, and its eponymous village. Visible throughout the fjord, the mountain serves as the point of reference for all travelers navigating by boat or by dog sledge.

A leading factor that contributed to the accelerated flow rate was loss of the glacier's outer edge, which was grounded on a bar at the terminus of the fjord and served as a plug, restricting the glacier's flow and discharge, until 2003 (Holland et al. 2008). Fig. 9.15 shows the extent of this ice shelf (or plug) in the 1930s. By the early 2000s the plug had disappeared and the flow rate had greatly accelerated. The ice receded five miles during the past seventy-five years, one human lifetime. One causal factor for increase in the glacier's recession may be hydrographic. Changes in North Atlantic Oscillation weakens a subpolar gyre, permitting warm waters from Irminger Sea west of Iceland to move west toward Greenland, where it mixes with the cold south-flowing East Greenland Current and then flows north along the West Greenland coast, warming the waters of the fjord and melting its ice from below.

UUMMANNAQ

The island of Uummannaq, located in Qaasuitsup Commune, is the commercial and service center for outlying settlements on Uummannaq Fjord and the northern half of the Nuussuaq Peninsula. Uummannaq has a human population of 2,348 and more than 6,000 dogs. Uummannaq, Nuuk, Sisimiut, and Ilulissat, constitute nearly half of the total population of Greenland. About four degrees north of the Arctic Circle, Uummannaq has blue sky and massive rock outcroppings, which remind me of New Mexico and Arizona. During the summer months there are 24 hours of direct sunlight.

Uummannaq has a compact harbor surrounded by brightly colored red, yellow, blue, green houses like those in other Greenlandic communities. Here houses are bolted to a steep grade of bedrock below

9.18 Living on the Edge
This Arctic village with its seemingly precarious perch on the land and its vibrantly painted houses, countering the lifeless glacially scoured grey rock, is characteristic of many small towns along Greenland's coast.

9.19 Lutheran Church
This view of the stone church at Uummannaq features another seasonal sign, the cheerful summer daisy.

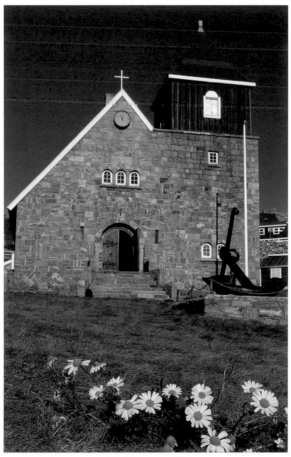

the island's spectacular heart-shaped peak, the geographic feature that gives the town its name. The community expands by blasting house sites out of solid rock, and each tier of homes is connected to the ones below and above with wooden staircases and walkways.

In May of 2004, the ice in Uummannaq Fjord was still solid, but it was too soft for travel, and there was not yet enough water for boat navigation. However, within three days the ice began opening up and boat travel began. The sun was up all day and hunters were racing around shooting seals to fill depleted larders. In August and September 2005 there was no pack ice and the few icebergs drifting about Uummannaq posed no threat to navigation. Ice conditions have become increasingly erratic on both sides of Baffin Bay. I traveled to Uummannaq three times between May 2009 and June 2010, primarily to dog sledge on the ice. But during none of these visits was there enough fast ice to make such travel possible. In early 2010, I planned to dog sledge and hike part of the new Greenland sections of the International Appalachian Trail. With no fast ice, travel by dog sledge was not possible, and our hike was limited to displaying the colors of Greenland at the newest IAT chapter. Throughout the winter of 2009–2010, there were only five weeks when travel on the ice was safe.

The German scientist Alfred Wegener—a world-famous meteorologist, geologist, and explorer who developed the theory of plate tectonics, also known as continental drift—was lost on a scientific expedition east of Uummannaq in 1930 (Kehrt 2013). Wegener pioneered the study of the Arctic as a cohesive circumpolar region and was the architect for the International Polar Year (1881–83). This study was unique in coordinating research teams around the edges of the Arctic to see if weather and auroral events could be understood as part of a single unified system or were isolated unique events; the results gave birth to modern meteorological science. Little is known about Wegener's disappearance, along with a Greenlandic companion (Vaughan 1994: 228), but he had reported to the Governor of Jakobshavn, Aage Knudsen, on September 11, 1930, that his team was having difficulty moving 100,000 kg (220,000 lbs) of supplies onto the icecap by horse, dog team, and propeller-sledge in preparation for the winter's work. Soggy snow had ruined the track and the party began to run out of hay for the horses. Wegener died less than two months later, in early November. I had learned of this story at the other end of the Far Northeast in Maine. Inger Knudsen Holm, Aage's daughter, shared her translation of Wegener's letter when she learned of my interest in Greenland.

Børnehjemmet, the "Children's Home," of Uummannaq, was founded in 1929, and has evolved into both a residence and an educational institution for troubled youth. The goal of the Children's Home is to teach young people to adapt to changing conditions without losing their Greenlander identity. The guiding principle is that patience, observation, learned skills, and rational thought—not anger—can provide solutions to life's problems.

Learning how to build houses of snow or wood, how to drive a dog sledge or traveling to Europe or the United States are intended to form a person's identity and to develop social skills for both community living and life away from Greenland. Foreign trips provide a frame for cultural reference. Most of Børnehjemmet's young people go to Denmark or elsewhere for further education and adaptation to western cultures. Teenagers work on language skills and learn appropriate behaviors of other cultures.

Experiential training is central to Børnehjemmet programs. Young people learn marketable skills like making traditional clothing or operating computers, along with retailing, painting, electrical and mechanical skills. Elders teach them hunting and fishing skills needed for living on the land. Each boy

9.20 Bow Hunting
On Uummannatsiaq, activities include fishing, hunting, target shooting, boating, hiking, and the joy of just running around. The boys have made their bows from PVC pipe.

9.21 Experiential Learning
Ane Nielsen prepares for a botanical field trip to classify berries, grasses, and other vegetation.

9.22 Dash of Color
Bright orange lichen frames this view of Uummannaq from Ikerasak Island.

9.23 Fishing
These Greenlandic boys are casting hand lines.

and girl experiences a dog sledge expedition on the winter ice, which is a rite of passage from childhood to young adulthood. On film, Children's Home staff record these young Greenlanders driving a sledge, caring for the dogs, setting up remote camps and fishing and hunting; these films become permanent records of the experiences of young people in surmounting their challenges.

Uummannatsiaq, "Little Uummannaq," is located on the northern tip of Ikerasak Island at the base of a mountain and a glacial corridor. The settlement rests on a glacial outwash plain, but the village was abandoned in 1969 when most of the population migrated to the settlement of Ikerasak and others to Uummannaq. When its resource base was reduced beyond a certain point, this settlement's population, like many others, moved elsewhere where there were family connections, seals, and fish. Subsequently, the four remaining buildings and the surrounding land of Uummannatsiaq were transferred to the stewardship of Børnehjemmet, which uses the facility to teach traditional Greenlandic ways of hunting and fishing. Amidst glacial striations, bright orange lichen, and tufts of grass with large seed heads, are the footprint of buildings that are long gone. Indentations in the land reveal where turf was removed to use as insulation for lumber-built structures. In the island's main settlement, also known as Ikerasak,

9.24 Sledge Dogs
Dogs pull sledges used for seal hunting or ice fishing, but when the ice is not safe and seals cannot be hunted, feeding dogs becomes a hardship. Unwanted dogs are destroyed periodically by officials designated for this unpleasant task.

at the southern end of the island, there are fine examples of turf houses and another field station of Børnehjemmet.

Ukkusissat, named for its soapstone deposits, which were used by the Inuit to make traditional lamps and pots, is located north of Uummannaq, near the head of the fjord and is the site of another Børnehjemmet field home. The settlement has the heavy capital accoutrements typical of a fish-based economy: large metal-hulled boats, a fish plant operated by Royal Greenland, and large fuel tanks. Its population seemed stable at 230 people in 2000, but since the Black Angel mine at Maamorilik, its economic mainstay, has closed, the population has decreased by about one-quarter, to 170 in 2010.

9.25 Qaarsut
The island of Uummannaq with its
heart-shaped peak is a dramatic sight
from Qaarsut.

EXPLORING UUMMANNAQ FJORD

On our first visit in 2002, Arne Nieman, the manager of the Uummannaq Hotel, made arrangements for us to travel on helicopter milk runs to four nearby island communities: Ukkusissat, Saattut, Qaarsut, Illorsuit, and Ikerasak. Each trip took us to a different settlement where we dropped off and picked up passengers and supplies. I photographed the magnificent landscape of mountains and snow, the sea with its icebergs and leads, and the cultural landscape of these outlying settlements and their people.

QAARSUT

Qaarsut, "the rocks," is located on the north side of the Nuussuaq Peninsula. Flying over Nuussuaq

Peninsula, one is impressed by eroded dikes of dark rock shaped like giant dinosaur teeth (fig. 2.05). These dikes formed when basaltic magma intruded into faults in the bedrock deep in the earth. After millions of years of uplifting and erosion, these rocks reached the surface, exposed as crisscrossing bands of dark rock. Here, the surrounding strata are softer than the magma and so the dikes stand alone, taking on strange shapes.

Lindsay had agreed to help teach English to the thirty-five students in the primary school at Qaarsut, which would permit me an extended period to roam the town and hills seeking photographic subjects. Arriving by plane, we were picked up by the town's only vehicle, a chattering van operated by the local

217

Pilersuisoq store, which functions as a taxi and carries anything else that needs transportation. We were dropped at the home of Esben Skytte Christiansen, our host and principal of the Kaalip Atuarfia (School), and discovered him to be a Dane with a working knowledge of five languages, an extensive library of European writers, and special expertise on Hans Christian Andersen. Esben's welcome was typical of the generous hospitality all Greenlanders provided us. In the evenings, after meals, we listened to the piano music of Erik Satie and watched icebergs floating by in the rosy evening light.

The monumental, transfixing presence of Uummannaq Mountain at 1,176 m (3,860 ft) governs the skyline. It was particularly brilliant in the early morning light, with red rays cast on the mountain and icebergs. Otherwise, Qaarsut is somewhat austere, but its starkness is moderated by the colorful array of Greenlandic homes. The population depends almost exclusively on the sea for livelihood, with hunting and fishing providing most of the food and a "skin factory," a sealskin apparel–manufacturing establishment that employs some local people and brings cash to the community. In 2004, Qaarsut had 205 residents and at least twice that many dogs. Throughout much of northern Greenland dogs are still the basic economic lifeline, which make occupation of Arctic regions possible. In the past losing one's dogs was tantamount to death, for without dog traction a hunter could not find sufficient food for his family within walking distance. Dogs also provide crucial

9.26 Greenlandic Teachers
Three hardy "pedagogues," the local term for the combined function of teacher, social worker, and psychologist, are kept warm and protected from wind and water by their sealskin parkas.

9.27 Arctic Fashion
Aaju Peter models sealskin clothing that her company manufactures in Iqaluit, Nunavut.

218

9.28 Qaarsut
Arctic flora and an iceberg frame the settlement of Qaarsut on the Nuussuaq Peninsula.

protection from polar bears, both as sentinels, and in helping humans hunt bears. Furthermore, they sniff out seal holes, are more alert to weak ice than humans, and can bring one safely home through a blinding storm even if the driver is incapacitated or unconscious.

The Greenland dog is genetically close to a wolf and exhibits wolf-pack behavior when in groups. Sledge dogs do not bark; they howl incessantly, creating waves of sound that ebb and flow across the settlement throughout the day. Each Qaarsut family has, in addition to free-ranging females and puppies, about a dozen dogs. Mature male dogs, by law, must be chained when in the village. During the summer dogs are allowed to run free on uninhabited islands, where they forage for themselves. Dogs are used exclusively for travel and hunting and are not considered pets; they create far less pollution than snow machines, automobiles or aircraft. Of course, dogs have to be fed, and this pursuit is a major part of the settlement's subsistence activity, probably equivalent to what is required for human consumption. Keeping and using dogs is a tradition Inuit have perfected since early Inuit peoples acquired them from Siberia about 1,500 years ago.

This land presents a paradox of nature. The same photographic composition may include an iceberg, with a glacial origin that may be measured in millennia, and the tiny woolly lousewort (*Pedicularis lanata*), with a single flower that blooms for perhaps a week on thread-like stems. Short-lived and long-lived is the nature of things. If I tread on one of these perennial Arctic blooms, it is gone for a year, if not forever. But one iceberg can sink a very large ship. This imbalance, this juxtaposition in space and time, reminds us that on this land, nature is still the master.

9.32 Greenland Meadow
This watercolor by Rockwell Kent includes both vegetation and ice, a duality that is the essence of Greenland.

primary commodities mined, but demand for rare earth minierals created the right market conditions. for reopening the mine.

ILLORSUIT

The settlement of Illorsuit, which means "big houses," is located at the north end of the island of Ubekendt, or Ubekendt Ejlande, which translates as "Unknown Island," from Ubekendt, Danish for "unknown," and Eijlande, Dutch for "island." Many places in Greenland retain names assigned during the seventeenth-century Dutch whaling period. Artist Rockwell Kent, who had first recognized his predilection for northern climes by living on Monhegan Island in Maine for five years, resided here in the early 1930s, during one of three extended periods in Greenland. (Kent also brought new attention to *Moby Dick* by illustrating whaling, and he lived for two years in Newfoundland.) I was sad to learn that dwellings

from his time, including his home, have been lost. But I made a photographic record of aspects of the scene that I felt Kent would have painted or illustrated, including buildings, fishing boats, glaciers, and mountains.

When we visited this settlement, the permafrost had melted into the muck of the Arctic tundra, which harbors resurrected dangers. Bacteria such as tetanus, once thawed, can again become infectious. A Swiss neurosurgeon in our traveling party worried that few precautions have been made against this threat—another potential implication of global warming.

9.33 Artist's house
In 1932 Rockwell Kent drew this illustration in pen and ink of his house in Illorsuit.

9.34 Illorsuit
Illorsuit, a small parcel of land precariously positioned just above the sea at foot of a mountain, is where artist Rockwell Kent (1882–1971) lived with Salamina, his Greenlandic paramour. Nothing remains of Kent's home now, but his drawing (opposite) from 1932 documents it precisely.

NUUGAATSIAQ

Nuugaatsiaq, the northernmost settlement in Uummannaq Fjord, is accessible by helicopter only twice a week, weather permitting. Located on an island, the settlement of Nuugaatsiaq, is limited to a spit of land that points south towards the settlement of Illorsuit, some 23mi/37km removed. Nuugaatsiaq is the most isolated settlement in Uummannaq Fjord. I stayed with the settlement's primary teacher, Barbara Strøm-Baris, an English-speaking German Dane, who reported on how

traditional hunting and fishing were being modified to cope with climate change. Barbara informed me that just that month her students had conducted a survey that established a demographic profile of the community. The population of 84 included 66 adults and 18 children, distributed among 29 families and 32 homes. There were equal number of boys and girls, but in the adult population, men outnumbered women, 35 to 30. Capital stock included 320 dogs, nine snowmobiles, and three ocean trawlers.

Ole Møller, a local hunter and fisherman, took me in his trawler *Helena* to hunt seals on the ice,

9.39 Nuugaatsiaq
The entire settlement of Nuugaatsiaq is poised on a narrow peninsula at the far northern reaches of Uummannaq Fjord.

which we found too thin and unsafe on our first attempt. When we located a floe solid enough to walk on, Ole stalked and shot a seal in the old traditional Greenlandic way, by crawling toward it behind a blind made of a small square of white canvas attached to his rifle. In the days that followed we spent most of our time fishing by long line and net for halibut or turbot (*Reinhardtius hippoglossoides*), saltwater catfish (*Anarhichas lupus*), and Greenland shark (*Somniosus microcephalus*).

Commercial fishing is the basis for the local cash economy. If travel by boat through the ice pack becomes too dangerous, the plant shuts down. The catfish that is not exported remains in the community for dog food, along with dried shark meat. Sharks have an unusual physiology: they do not have kidneys; rather, they excrete urine through their skin. Because of this, shark must be processed by drying it on racks high enough to keep it from dogs. If dogs ingest the meat or skin before it has

completely dried, they are likely to be poisoned by the high concentration of ammonia.

Other wage-paying positions are in teaching, medical care, power plant operation, a service center, the community office, and Pilersuisoq store. In Nuugaatsiaq there are an estimated three full-time positions and ten seasonal or part-time positions; all wage-paying jobs, except those in the Royal Greenland fish plant, are with the central government or municipality.

Formal education is not seen as being particularly relevant, and there is little interest in learning either Danish or English. There is strong family pressure for young women to marry so that elders can have grandchildren. Conversely, young women are more likely than young men to see the value of education and subsequently leave the settlement. Young people must go to Uummannaq to complete the required level of education (tenth grade), but they often drop out and return to continue life in settlement. The

9.40 Hunting Success
Hidden behind a white blind attached to his rifle, Ole Møller moved stealthily forward and took this ringed seal (*Puda hispida*) with a single shot.

9.41 Waste Not
Hunters flense a ringed seal, separating the meat that will be used for food from the hide that will be crafted into winter boots.

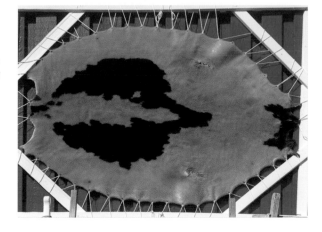

9.42 Harp Seal
The pattern on harp seal's back resembles a map of Greenland. One wonders if this played a role in its name, *Phoca groenlandica,* or whether it is because this species is particularly numerous and important in the Greenland economy.

traditional subsistence requirements of that life are increasingly transformed by climate change and are unpredictable. New conditions that arise because of climate warming and sea-ice reduction will require changes in technology and hunting and fishing strategies: larger seagoing boats and more outboard motors. Snowmobiles are another tool, but they are expensive to buy, rely on fossil fuels, and are only as dependable as the ice on which they are driven for hunting and fishing.

In Greenland, new economic ventures include agriculture, which after a multicentury gap is now a major employer; both root crops and livestock are profitable. Mining, too, is taking hold in those sections of Greenland that are no longer ice covered. The tourist trade has stimulated craft production: artisans are refining techniques for using musk-ox hair in woven shawls, hats, and other clothing. Increasingly, objects are being carved from reindeer antler, and even from the hoof pads of the animal.

DARK TIME IN THE ARCTIC

Our trip to Greenland in 2009 was the first time Lindsay or I experienced an Arctic winter solstice. In Maine we traditionally celebrate the winter solstice with neighbors around a bonfire in the woods behind our home, and then by the fire with food in

9.43 Half Moon
With very little cloud cover, light reflecting off the waters of the fjord is supplemented by moonlight.

9.44 Uummannaq at Christmas
Although Uummannaq is north of the Arctic Circle, in winter, light from below the horizon is refracted above the horizon for about four hours per day, producing twilight illumination.

9.45 Christmas Dinner
Guests at Christmas dinner pose with hostess Ann Andreasen (right). I was as tired as I look, not being accustomed to so much "dark time" socializing.

9.46 Winter Streets
Greenlanders walk to Uummannaq's center against a bitterly cold wind, down streets illuminated by house lights that keep the encroaching darkness of an Arctic night at bay.

9.47 Christmas Eve Services
In 2010, Lindsay and I attended Christmas Eve services at the Lutheran church where the choir in traditional dress sang European hymns in Danish and Greenlandic.

our home. We were concerned about our ability to adapt to the months-long period of Arctic darkness. On our flight from Copenhagen to Kangerlussuaq, the exterior temperature was −84 degrees F, and it seemed my concern would become reality. But by the time we reached Uummannaq, the situation had changed: there was some light, temperatures were above what we might expect in Maine, and Greenlandic hospitality was abundant and genuine. We found winter in Greenland filled with social warmth replete with friends, dinner parties, and *kaffi mik*—coffee and cake receptions. At home in Georgetown, we enjoy the amity of small-town life. Even though we speak neither Greenlandic nor Danish, we experienced much the same *communitas* in Greenland. Our hosts Ann Andreasen and Ole Jørgen Hammeken kept us busy with dinners at their home, with their friends, or at the Children's Home. Bountiful meals ranged across the local ecological spectrum: from narwhal,

9.48 Cool (e) motion
On March 22, 2010, a sculpture created by Dutch artist Ap Verheggen was placed by helicopter on an iceberg in Uummannaq Fjord. Its placement on a melting iceberg that would inevitably destroy the work expressed the artist's visceral reaction of climate change.

halibut, seal, Arctic char, shrimp, and caviar to reindeer, musk ox, and ptarmigan—all accompanied by a good complement of Danish food as well as fruit and nut pastries, sweet butter, tapioca-based whipped cream, and drinks including Carlsberg beer, Gammel Dansk, and a selection of wines from the Pilersuisoq store.

Although Uummannaq at 71° North is just short of five degrees north of the Arctic Circle, the temperature was often above freezing—not the harsh cold we expected. And the noontime sky is still somewhat light. Layered bands of light, ranging from clear white to pastel red, emanate from below the horizon and reflect off the atmosphere, washing over the village from about 11:00 am to 2:00 pm.

On Christmas Day it was so bright for a little while that it seemed like a spring day. The sky became overcast for a few days after Christmas, reducing much of the detail of the village and surrounding land, rendering all my outdoor photos a pale blue. Winds were strong, knocking out both television and internet transmissions. At the end of December, the temperature was 6°C (43°F), warmer than much of Europe or North America. On the last day of the year, when the temperature is usually around –20 or –30°C, a strong, warm wind blew from the south. In Uummannaq Harbor, fast ice was melting into expanding pools of water, acting as mirrors reflecting the ruby light from the horizon. And on the last day of 2009, there was a torrential downpour in Nuuk and hurricane-force winds were forecast to the north, in the town of Upernavik.

During much of the day artificial light streaming from windows and the exterior decorations of homes makes up for the dearth of natural sunlight. Light cuts through the Arctic night in the form of violet rays from street lamps and the red, white, and green Christmas decorations. In the village center, a solitary Christmas tree stood next to the Lutheran church, resplendent in white lights, each one with its aura of a full moon. In the deepest, darkest space that only an Arctic night can provide the northern lights shattered the cosmic darkness. But in Uummannaq around December 27, the "northern lights" are more likely to be a pyrotechnic display of fireworks that are part of an annual event. In Uummannaq, we experienced a hybridized Scandinavian/Greenland culture that holds a bright light to the Arctic night.

9.49 Midnight Sun
In June, even though the sun casts a golden glow in the middle of night, the air remains cold and, after a few hours on the water, drains one's energy. After travel in these conditions, cognac and chocolate are a suitable antidote!

CHASING AN ICEBERG

On March 22, 2010, a sculpture of two dog sledges made of armored electrical cable was placed by helicopter on an iceberg in Uummannaq Fjord. Produced by Dutch artist Ap Verheggen and entitled *Cool (e)motion*, this composition of metal and ice and its placement reflected the artist's view of what is happening because of climate change, in contrast to scientists' investigations of the why and how of the phenomenon. Putting a dog sledge sculpture on iceberg was Verghen's way of saying that nature is in charge of human affairs, not the other way around. In other words, nature will prevail—in this case through warming of the planet.

Cool (e)motion was also intended to raise public awareness about the effects of climate change on people and culture. Verhaggen had arranged with Google Earth to monitor the movement of the iceberg on an hourly basis and provided an opportunity for people around the world to comment on climate change from their corner of the planet. The intention was to create a citizen-based statement on climate change, which could help leverage political action to address the problem. The undertaking was elaborate. However, the iceberg broke up without ever leaving Uummannaq Fjord. Although the world missed the opportunity to follow what could have been a long and curious story, the berg's rapid destruction and the premature loss of the sculpture told its own tale about the effects of global warming.

UPERNAVIK

Upernavik, "the spring place," and its outlying settlements have been folded into Qaasuitsup Commune. The town has about 1,200 residents, and with surrounding settlements, numbers 2,834. Upernavik (N72° 47'; W56° 09') is located about 40 km (24 mi) from the outer coast in a group of islands and is at almost the same latitude as Pond Inlet (N72° 44'; W77° 21') in northern Baffin Island. Until active polar exploration began in the late nineteenth century, Upernavik was the frontier of Danish Greenland. Its northern location and limited accessibility by plane has extended its early history of isolation and frontier ambience.

The Greenlandic Norse first reached Upernavik in the thirteenth century. In 1824 hunters found a small hand-size stone tablet with a runic inscription in a rock cairn on the island of Kingittorsuaq about 20 km (12 mi) northwest of Upernavik. The inscription reads: "Erlingur Sigvatsson, Bjarni Thordarson, and Enridi Oddesson built Saturday before Rogation Day a cairn" and is dated 1323 (Gulløv 2000: 321). Other than archaeological finds of Norse artifacts in Thule sites in northwest Greenland and nearby Ellesmere Island—which could have arrived there by trade or been scavenged from Norse ships or won in battle—this small runestone is the northernmost direct evidence of Norse activities in Greenland and the only authentic runestone known in the Americas.

South of the business center is the historical district, which consists of a museum, an old shop, the old church, and the bakery and cooperage, now used as a residence for visiting artists. Nearby is the grave of Peter Freuchen's wife, Navarana, who was then known by her Inuit name, Mekupaluk. She had accompanied Freuchen and Knud Rasmussen on their early Thule expeditions and bore two of Freuchen's children. Freuchen buried her here on a ledge overlooking the sea and the church whose minister had forbade her burial on consecrated ground in the churchyard because she had not been baptized a Christian. Her grave has become a pilgrimage site for indigenous people as well as tourists.

REFLECTION

Travel makes one of aware of the importance of being connected to a place, be it settlement, village, town, city. I have come to believe that a path to wisdom is possible through life lived out on the land, particularly in challenging environments. Others have recognized the same truth in visiting Greenland: "When you have been with those people —with the Inuit—you know that you have been with human beings" (Ehrlich 2001: 277).

9.50 Kingittorsuaq Runestone (Replica) Greenland Inuit found a small runestone in a cairn on Kingittorsuaq, north of Disko Bay. The inscription tells of three Norse hunters who visited this spot at the end of the thirteenth century.

9.51 Merging Rivers of Ice
Trim lines and moraines along the sides of these mountains reveal higher levels of glacier ice that have since receded with warmer modern conditions. Dark bands of eroded debris merge as each ice tributary moves toward its outlet.

235

9.52 Imaginative Play
This young fashionista is playing *ajarraaq*, a cats-cradle-like game that includes string figures for different animals, birds, and people.

9.53 Soccer Enthusiasts
Girls at a soccer game in the town of Itilleq, Greenland, cheer on their team.

9.54 Imagining Another Life
Lindsay wearing traditional Greenlandic woman's clothing to a Christmas celebration.

9.55 Girls in Full Fashion
From childhood through adulthood, Greenlandic females wear this striking traditional costume for festivals, important social events, and national holidays.

Aqqaluk Lynge, Greenlandic writer, poet, President of the Inuit Circumpolar Council and President of the newly formed Center for Arctic Human Rights, writes of the struggle for cultural survival facing the indigenous peoples of all lands of the circumpolar north. In a poem "The Wind From the South" (2008: 113), Lynge writes that the Inuit have historically submitted to the West, but no longer:

> For too long we have turned our backs against it [western world]
>
> For too long we have bowed down in respect
> For too long we have treated the wind
> With a dignity it does not deserve.

I try to remember that when I am in the North, I am there as a student of cultures and peoples who have survived the harshest of climates and most demanding circumstances. After two centuries of Danish administration, Greenlanders are beginning to exercise self-determination within the context of nationhood.

Indeed, the future of the Arctic is not what we once thought it would be. Who would have expected the Arctic Ocean to become seasonally open or this formerly marginal region to begin playing a pivotal role in the history of the world? These changes, as well as those adrift in the winds of the south, including recognition of climate change and the need to limit growth, will shape the future of the entire planet.

9.56 Girl in Window
The settlement of Nuugaatsiaq is so small, existing on a tiny peninsula, that strangers, like this little girl, enter our lives easily.

Journey's End

At the conclusion of our virtual journey, we feel compelled to reflect on these northern lands and the future of its peoples. The Maritime Far Northeast has not figured prominently in the calculus of nations since Martin Frobisher initiated the search for the Northwest Passage in the 1570s (see Chapter 7). Many of the people living in this huge pristine region continue lives and cultures that have existed for thousands of years. Our wanderings through this landscape—dipping into its history and cultures and its episodic interactions with the rest of the world—have revealed a place that largely for reasons of geography and climate has stood apart—grazed by whalers, occasionally exploited by miners and fishermen, and coerced into hosting Cold War Distant Early Warning and military installations—while the rest of the world has raced forward.

The Maritime Far Northeast is now experiencing an unprecedented political, cultural, and economic awakening that will involve the rest of the world, which will soon to be exploring for oil and minerals and sending ships through the Arctic Ocean and the ice-free Northwest Passage. Our purpose in writing this book has been to bring greater public awareness to a long-neglected region that is destined to play a larger world role in the future than it has in the past. Climate change, resource extraction, commercial ocean transit, environmental pollution, and the future of this northernmost region's native populations, are key issues for the North as they are also for the wider world. A rational way forward is needed; in turn, the small-footprint societies of the Maritime Far Northeast may provide models of adaptation useful for sustainable development in the rest of the world.

CLIMATE AND ADAPTATION

The geographic chapters of this book have documented the many similarities, continuities, and connections that exist throughout the Maritime Far Northeast. Most are rooted in the region's dominant physical characteristics: cold climate, cold waters, and an ocean-current system that exports nutrient-rich waters as far south as Nova Scotia and sea ice to northern Nova Scotia. The shores linked by the East Greenland, West Greenland, and Labrador currents share a common host of sea mammal, sea bird, and fish species. Land resources were more geographically diverse. The most important of these land and sea resources—caribou, moose, harp seals, bowhead whales, salmon, and codfish—provided food, hides for boat covers and clothing, and building materials. These species provided a common base for the region's cultures, allowing for those with suitable adaptations to populate the region. Eskimo cultures could easily spread from Greenland far south along Subarctic coasts, while Indian tribes sharing closely related Algonquian languages were able to spread as far north as the range of the interior-dwelling woodland caribou and other forest animals, from Maine to Labrador and Quebec.

There are many lessons to be learned from the history and adaptations of northern cultures. One of the first is that change is inevitable and the ability to adapt or to mitigate the threat (for instance by reducing CO_2 emissions) is key to survival. Humans cannot irrevocably force nature to their bidding, especially in the Arctic, and cannot always rely on technological innovations to prevail against adversity. Eastern Arctic prehistory is full of such lessons. The Pre-Dorset and Thule migrations into the eastern Arctic were opportunistic gambits that were facilitated by warming events. Pre-Dorset people carried an Arctic land-hunting economy developed in Siberia into what is now Alaska, Canada, and Greenland, but once there had to diversify and adopt toggling harpoons for sea-mammal hunting to secure survival when faced with caribou-population crashes. Dorset people who hunted seals and walrus at polynias (areas of open-water caused by wind or upwelling in the midst of dense pack ice) in the High Arctic were faced with local extinction or southward migration when these game "oases" froze over during the cold Sub-boreal Period (500 BC–AD 500). Later, during the Medieval Warm Period (ca. AD 800–1350), Thule culture peoples discovered a new niche for their whale-hunting economy in the ice-free passages of the Canadian Arctic. Concurrently, a less stormy, more ice-free

North Atlantic encouraged Norse agriculturalists to settle in Iceland and Greenland. But once in Greenland, where their population quickly grew to about 5,000, the onset of the Little Ice Age (AD 1450–1850) forced retrenchment and eventual abandonment of this locale. Colder conditions encouraged Inuit sea-mammal hunters to reoccupy vacated Norse homelands.

The lessons are clear: climate change provides opportunities, and new technologies may also be

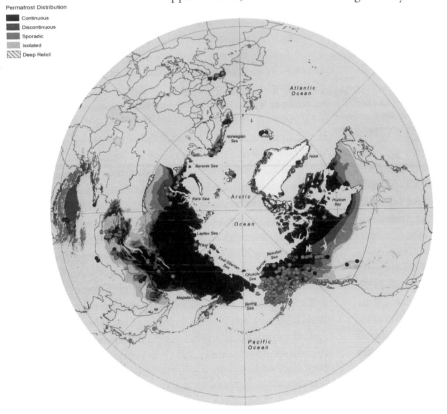

Permafrost Distribution
- Continuous
- Discontinuous
- Sporadic
- Isolated
- Deep Relict

10.01 Permafrost, Perception, and the Circumpolar World
The landmasses around the Arctic Ocean presents a strikingly different perspective of the world than the customary Mercator projection that has dominated geographic education. This map shows the distribution of permafrost, which—like sea ice—is shrinking and retreating to the north. Our new ability to virtually turn the world on its head reminds us that, as the physical and biological world responds to climate change, our perception of the world must also change. Nowhere is this more evident than in the Arctic.

required, but societies must beware of depleted environmental capital; cycles exist and good times become bad times. When human population growth intersects declining resources, disaster forces extinction, migration, or new subsistence strategies. Arctic hunting cultures and marginal agriculturalists like the Greenland Norse are excellent examples of the propensity for human systems to collapse. In this sense humans operate like most other biological systems. Often when one culture declines, another, with different subsistence strategies, benefits. These processes are clearly evident throughout the 4,500-year history of the Maritime Far Northeast. Most northern societies in the prehistoric record appear to have lived well within the sustainable band of the resource oscillation curve.

Climate matters. But until recently its role has not been fully understood or appreciated. The changing aspect of climate that humans react to on a daily basis is weather. The more significant changes in climate have a longer periodicity and more lasting impacts. During the Late Archaic period (5,000–3,500 years ago) in Maine and the southern Canadian Maritime provinces, oak was a major forest component; today these forests are mixed pine, spruce and hardwoods with little oak, but oak and southern trees are beginning to move north again, and white pines are displaying record yearly growth spurts. Climate-sensitive animals, insects, and plants are moving north; and people—and modern cultures—are too. According to Serge Payette, the northern Quebec boreal forest landscape that he studies "is changing in front of [our] eyes. The annual mean temperature in the area east of Hudson Bay has increased by at least 2°C [3.6 °F]. The permafrost is thawing, and the tree line has begun a slow march north. Along the eastern coast of Hudson Bay and Labrador, new seedlings have begun to take root beyond their historical limits" (*Canadian Geographic* 2012). A comprehensive analysis of world climate, based on ice-core analysis, documents the past century as the warmest in the last 4,000 years and projects that if current trends continue, the era since the Industrial Revolution will be the warmest of the entire 10,000-year Holocene postglacial period (Marcott et al. 2013).

Remarkable changes are also taking place in the marine system. For the past several years sea ice has been diminishing in the Gulf of St. Lawrence and around Newfoundland; open water has been occurring regularly during midwinter in the Disko Bay region of West Greenland and other areas, including the Northwest Passage. The 2012 NOAA Arctic Report Card recorded a host of new developments across the Arctic: increased size and incidence of meltwater ponds on the pack ice that render the ice transparent to sunlight have stimulated massive under-ice blooms of algae and plankton; record-breaking melts on the Greenland Ice Cap; Arctic Ocean sea ice continuing to thin and reaching a summer minimum 18 percent below its 2007 record (see Chapter 2). From the Arctic to Antarctic, from the Himalayas to Andes, glaciers are disappearing today. Humans have ignored these signs in the past: Bolivia's 18,000 year-old Chacaltaya Glacier, which began losing mass in 1940, was projected in 1998 to have disappeared completely by 2015. Without any fanfare, it disappeared sometime in 2009 because the rate of thaw had tripled since 2000 (Feldman 2009).

What requires more careful analysis is the long-term effect that these statistics portend for the environment and humans. The warming trends and

degradation that has come with outside intervention in its fisheries, forestry, and mineral developments.

In the Maritime Far Northeast, history is coming full circle. Through the early twentieth century, whales were sought for their oil, until technology replaced the diminishing number of whales in the earth's oceans with abundant supplies of crude oil extracted from the earth's crust. Now, as land sources are proving inadequate to meet petroleum needs, commercial interest has returned to the waters of the Far Northeast to procure oil from the Grand Banks of Newfoundland and to tap natural gas from the waters of Nova Scotia. Over the last few years, the seafloor lying beneath the icy waters of Nunavut, the Maritimes, and Greenland continues to be explored for oil, and discoveries are being made.

Energy is "the lifeblood of modern Arctic settlements," according to Julian Dowdeswell and Michael Hambrey (2002: 189), who go on to note that, "every mammal needs to generate its own heat, and this is 'expensive,' so it [a mammal] must be well-insulated to prevent heat loss" (2002: 195). From insects to humans, the importance of energy to maintain body warmth becomes a more critical condition of life as one proceeds toward the north. Current demand demonstrates that humans living in Arctic settlements are increasing their need for energy. Over the last century, per capita energy consumption increased four-fold while human population increased six-fold, and fossil-fuel consumption increased sixteen-fold (Flannery 2005: 77). These energy needs are real, but in both the Arctic and in our temperate world, there are sufficient alternate resources. People who utilize local resources and resource sharing to meet their needs are more economical. In northern forest regions many people are again heating mostly with wood. Iceland's 250,000 people are heating their homes and powering their industries with volcanic energy tapped from the locally thin crust of the Mid-Atlantic Ridge. Greenland has tied its future to electricity generated from glacial meltwater.

Environmental and social destruction does not need to be a subtext for progress, which can coexist with environmental and social conservation, if certain parameters are understood and respected. The Maritime Far Northeast can support only small populations: that is the key to their sustainable

10.04 Dorset Figurine
This soapstone figure from Shuldham Island, Saglek, Labrador, probably of a woman, displays the high-collared parka hood that is a distinctive, although seemingly impractical, feature of Dorset-culture clothing.

lifeways and approach to husbanding the land and its resources. A new phase of geographic expansion into the Arctic that coincides with increasing climatic and environmental stress threatens that balance.

Recently climate change has become an even more powerful force of change than the pressures of economic development, modernization, westernization and their attendant social ills. Rising sea levels are flooding out Arctic villages on Alaska's coast; melting sea ice is depriving people of their livelihood; and the world's demands for petrochemical energy threaten to endanger subsistence food supplies and the ecological web on which life depends. The international economist Jeffrey Sachs warns that while "science and technology can be harnessed…[T]he harder question is whether we will be well enough organized, and cooperative enough on a global scale, to seize the chance to save the planet from climate disaster" (2008: 74).

Robert Reich reminds us that humans must be more than just consumers and investors; we must be citizens of the global polity (2007). Being a citizen of anything requires a duality of rights and responsibilities. In the throes of rapid development, much of the Western and Eastern worlds have shrugged off their responsibilities as citizens and have chosen, rather, to be characterized as "consumers," a passive identity solely bestowed by the market. Such a categorization erases our sense of place, increasing the distance between people and land. A consumer is simply an economic category that excludes most of the other social, cultural, and spiritual aspects of human beings. As consumers, we are modern-day nomads ranging across the land seeking work and access to goods and services, wherever they can be found—in box stores, fast-food franchises, or the Internet. In Greenland, subsistence hunting-and-gathering cultures have persevered for at least 4,500 years. Yet, when these roving hunters are relocated to permanent settlements, either voluntarily or involuntarily, microcosms of suburban consumption patterns develop. Northern peoples in these growing towns and cities soon emulate the values of their southern relatives: more consumer goods and less connection with the land.

Two decades ago Native American writer William Least Heat-Moon expressed the rising consciousness of conflict between humanity and environment. Conceding that white men have become stewards of the land, he cautioned that these white men have "the big machine but not the operating instructions…the white man's…way of life is the land's death" (1991:

239). For two centuries, aboriginal peoples have been forced to emulate and assimilate the Euro-American model of progress through technological development and the expansion of the global economy. Today there are signs that the world is beginning to reverse direction and heed the advice and sustainable ways of indigenous peoples and individualists who live at

10.05 **Migration: Umiak with Spirit Figures**
Abraham Anghik Ruben's sculpture was exhibited at the National Museum of the American Indian in conjunction with the Arctic Studies Conference hosted by the Smithsonian in 2012. This work evokes the ancient traditions of his Inuvialuit ancestry of the western Canadian Arctic, where seasonal migrations marked the passage of time and relationships between animals and humans were mediated by hunting ritual and shamanic ceremony.

the tail ends of the bell curve. While the historical and modern societies of the North should not be idealized, they do provide some alternatives that could be adapted as models for sustainable living elsewhere. But, instead, over the last few decades, traditional aboriginal technology has been replaced by systems that dramatically increase the dependence on imported goods and petroleum products. As the North adopts the Western model it sacrifices its self-sufficiency and independence, the qualities that helped its peoples to adapt to changing conditions over thousands of

years. Sharing the natural wonders of the Maritime Far Northeast from Maine to Greenland is one way to raise consciousness of what we are in danger of losing, and to suggest what we must do to save it.

READING RUNES

In a cairn on an island northwest of Upernavik the small, palm-sized Kingittorsuak Runestone, chiseled by Norse hunters in the late thirteenth century, was found by a Greenland Inuit in 1824. Among today's "runes" are scientific reports about shifting climate; in a changed world that is much larger than it was in Norse times, these are the messages that must be decoded.

When Will's great-great-great-grandfather Moyse Richard was born in Lower Canada in 1800, the world's human population had just reached one billion for the first time. The US Census and the United Nations project that by 2015 our population will number eight billion; just during the lifespan of one generation, Earth's population will have quadrupled. Not only are there are more of us, we also live longer and have dramatically intensified our demands on the environment. The majority now resides in metropolitan spaces, and many of us live in a megacity—one with a population of at least 10 million—because the number of these urban areas has expanded from three in 1975 to twenty or more in 2009. Humans have been exponentially degrading the Earth's carrying capacity, for example, through the fastest rate of species extinction in the last 65 million years. The cause of this observed extinction is displacement of species habitat for the purpose of human food production, transportation infrastructure, and urban sprawl.

Four decades ago, in the oft-cited Club of Rome's *The Limits to Growth*, its authors (Meadows et al. 1972) gave new life to the eighteenth-century Malthusian doctrine that population growth outdistances the ability of technology and the planet to support the rising numbers of human beings. Now, in the second decade of the twenty-first century the demands of seven billion people may be causing the earth to approach a tipping point beyond which there is no return for this tiny planet floating in the void of space. As economist Jeffrey Sachs writes in *The Price of Civilization*: "[The] human impacts on nature are for the first time in human history so great that they threaten the planet's core biophysical functioning.... Our peril is unprecedented, and human knowledge, values and social institutions are far behind the curve"

243

(2011: 175). The paradigms of economics and ecology must coalesce into one, and reconcile the true costs and benefits to humanity and to the land that sustains us.

Since the Renaissance naturalists and academics have explored the world, discovering and dissecting its constituent parts and building systems of thought that described and explained them, creating scholarly disciplines such as medicine, physics, biology, electronics, anthropology, and many others. Today we are discovering that what once seemed like rigid boundaries between fields are porous and are interconnected in complex ways. Disciplines are now merging into sets of systems linked by interdisciplines: ecosystems, bioregions, molecular biology, and biochemistry, string theory, political economy—perhaps all eventually to be incorporated into some grand theory or set of linked systems. Increasingly, economics is integrating costs of production (air and water pollution) into the cost of doing business—perhaps along lines that E.O. Wilson would approve. Decades in the making, economics and ecology are becoming increasingly connected and, in fact, are cooperating in ways that are encouraged or demanded by governments.

From the activities of individuals to heads of state, new ideas are taking root as the world awakens to the effects of climate change and the concerns of northern peoples. Among the important new connections between these lands in recent decades is the political integration resulting from pan-Inuit political organizations. The new governments of Greenland and neighboring Nunavut have given their Inuit populations a greater role in establishing policies that in many respects lead the world in areas of cultural and environmental awareness and sustainability. Informed by a northern indigenous perspective, the Inuit Circumpolar Council has brought Inuit cultural values to the floor of the United Nations, influencing the policies of nations and sensitizing world leaders to the need for meaningful consultation with northern people. The newly created Arctic Council, with representation from Arctic nations, has quickly established its legitimacy as a pan-Arctic political organization and has contributed to the formulation and coordination of international policies such as trans-Arctic shipping, search and rescue, pollution, and industrial development. These political developments are being spurred by the crisis of climate change currently being experienced in the North.

The attention of an international group of scientists, researchers and museums has produced some of the most useful and well-grounded information. Climate studies—which have included four decades of collaboration between the National Science Foundation and Danish scientists on the Greenland Ice Sheet Project (GISP-1, GISP-2); glaciology studies on Baffin and Ellesmere Islands by the universities of Colorado and Massachusetts; and multinational studies of sea-sediment cores in Baffin Bay, the Labrador Sea, and East Greenland Sea by the Canadian Bedford Institute of Oceanography—have produced many pioneering discoveries. Early on the GISP collaborators learned that air bubbles trapped in ancient glacial ice constitute an archive of the earth's atmospheric "pulse" that could be tracked back hundreds of thousands of years (Oeschger et al. 1985). The ice sheet data has yielded information on changing CO_2 levels, documented the history of volcanism, and established a record of storminess—which coincides with more sea salt being transferred into the atmosphere (Mayewski and White 2002). Perhaps the most striking revelation from Greenland's ice cores was that global climates have shifted from cold (glacial-dominated) to warm in a few decades or less—which is almost instantaneous on the geologic time-scale (Alley 2002). These and other projects have produced hundreds of scientists who have worked across the borders of the MFNE, and this pattern is now expanding to include Canadian, American, Greenlandic, and Scandinavian students who frequently seek higher education in each other's countries.

Scholarly and educational exchange programs have also blossomed. The Smithsonian built upon its long history of archaeological research ranging from the Gulf of St. Lawrence to Labrador and Baffin Island to explore one of the more importat Nordic–North American immigrant legacies, Leif Eriksson's Vinland voyages, in a major exhibition and book, *Vikings: the North Atlantic Saga* (Fitzhugh and Ward 2000). The Nordic countries, Greenland, and Canada cooperated with the Smithsonian in producing this exhibition that also explored the growing archaeological evidence of Norse contacts with Canada's Dorset, Thule, and prehistoric Indian populations.

In May 2005 the Smithsonian Arctic Studies Center mounted a *Festival of Greenland,* cosponsored with the Home Rule Government of Greenland and the Embassy of Denmark. A large contingent of Greenlandic crafts people, musicians, and government representatives were present, including Robert Peary IV, the great-grandson of Robert E. Peary and the Inughuit (Polar Inuit) woman, Aleqasina. Other connections between Maine, Labrador, and Greenland have continued through the legacies of Peary and MacMillan and the modern research programs of Bowdoin College's Peary–MacMillan Arctic Museum.

In October 2012 the Smithsonian Institution's Arctic Studies Center hosted the 18th biennial Inuit

10.06 Gannet Aloft
Gannets, one of the most majestic large birds of the northwest Atlantic coast, have been constant companions as we cruise the Maritime Far Northeast. Unfortunately we have lost to human predation its former avian partner, the great auk.

Studies Conference, which attracted 600 participants, including scholars and Inuit from Canada and Greenland. Its theme, "Learning from the Top of the World," was enunciated in films, seminars, scholarly sessions, and exhibitions that brought indigenous representatives, Inuit youths and elders together with political leaders, policy experts, and government officials to discuss the impacts of climate change and ways that its deleterious effects might be minimized. Inuit youth expressed concern about rising sea levels, more powerful storms, mineral development and industrialization, pollution, and the social consequences of the immigration and globalization on Inuit language and culture. A highlight of the conference was a youth group from Uummannaq, northern Greenland, who performed classical and traditional Inuit music and presided over the North American premier of *Inuk*, a film about loss and affirmation of cultural identity among Greenland's young people.

THE CHALLENGE

The challenge we face is therefore to find some means of enjoying the same degree of harmony and tranquility as those of more traditional communities while benefiting fully from the material developments of the world as we find it at the dawn of a new millennium.

His Holiness The Dalai Lama, *Ethics for the New Millennium*

The return of the eagle and the puffin in New England and the Maritimes, restoration of whale species indigenous to the waters of the Maritime Far Northeast, the imposition of quotas on narwhal and polar bears in Canada and Greenland, and an increase in number of parklands throughout the region are all examples of successful stewardship policies. It is important that we preserve these living archives of nature, not only for humans but for the sake of this total ecosphere. Nature remains a repository of the sacred, and flourishing small-scale cultures communicate the sanctity of nature that sustains us in our built-up world.

As geographers, photographers, and scientists, and as concerned citizens, we and others are sounding the alarm of climate change and offer proactive small-scale solutions, many drawn from life in the Maritime Far Northeast. With climate change now convincingly documented and recognized as a worldwide phenomenon, we need to think beyond geopolitical borders. With a global market, international corporations, mass culture, satellite-based communications, English as a dominant language, and global warming, borders are passé except as pertains to issues of security. The environmental and social problems that we face are not contained within national borders.

Much like the genetic information contained within species of wild wheat or wild maize as a repository of benefits and costs, the knowledge of small indigenous cultures, such as those of the Maritime Far Northeast, can assist the global culture to find ways of living responsibly within that nature of which we are all very much a part. Like the transformation of a wooden or stone shelter into a home, it is only through intimacy with the land that we develop a sense of place. All too often, peripatetic ways and omnipresent technology act as firewalls to bar us from the world of nature and from ourselves. In these journeys the authors have never really left home: we have simply enlarged the circumference of home to include the wider reaches of the Maritime Far Northeast. We hope that these journeys will encourage you the reader to make the Maritime Far Northeast part of your world, too.

ACKNOWLEDGMENTS

THIS BOOK REPRESENTS A DECADE of collaboration between the authors but has been enriched by decades of earlier experiences from Maine to Greenland.

Over the years Bill's travels have brought him into the lives of countless individuals and families in Quebec, Newfoundland, Labrador, Baffin Island, and Greenland. More than just friends, people like trappers Henry Blake and John Michelin of North West River; Steve Tooktoshina, Elizabeth and Joe Goudie of Goose Bay; Bert Allen and Charlie Tooktoshina of Rigolet—all from Hamilton Inlet, Labrador—shared insights and knowledge in the 1960s and 1970s that guided my understanding of lives, landscapes, peoples, and cultures of Labrador. During the four years that I researched the voyages of Martin Frobisher and Inuit-European interactions in Frobisher Bay my work was facilitated by the Iqaluit Research Center, Mary Ellen Thomas of Arctic College, Paul Pishuktie, and many others to whom I owe a deep gratitude. At the more recent end of the time scale, Perry and Louise Colbourne and their extended family of Lushes Bight brought four hundred years of Newfoundland history and culture into perspective, as did the Evans, Rowsells, Vatchers, and others from the Quebec Lower North Shore. Clifford and Florence Hart of Brador shared their archaeological collections and entertained us with fresh baking and old-time fiddle music. These and many others showed us hospitality and educated us into the ways of the North.

None of this work would have been possible without the support of my wife, Lynne Fitzhugh, Stearns Morse who endowed the Smithsonian with Research Vessel *Pitsiulak*, Perry Colbourne who skippered *Pitsiulak* for many years, and the many students and colleagues who participated in the Smithsonian's northern research programs.

Will has had similar formative experiences. I owe much of my understanding and profound thanks to many: from my International Appalachian Trail colleagues and Greg Wood, who with me pioneered northern ecotourism in Quebec and Labrador to Boyce Roberts of Quirpon, who shared his deep knowledge of Newfoundland outport life and boating skills in ice-clogged waters, and Robert Bryan, Founder of the Quebec-Labrador Foundation, who encouraged and supported the compilation and publication of this book. Through a blend of curiosity and happenstance, and with the encouragement of my wife Lindsay, my northern travels continue. I owe a debt of gratitude to people I met along the way for their generous hospitality and for sharing their knowledge and experiences in the Maritime Far Northeast. In the mid-1990s, Richard Anderson had the idea of building an international version of the Appalachian Trail to bring people together across national borders by footpaths, initially from Maine's Mt. Katahdin to northern Newfoundland. Now, the International Appalachian Trail (IAT) has been extended onto Greenland, Iceland, Scotland, England, Wales, and Scandinavia, as well as elsewhere on the continent and islands of Europe.

For air flights to Greenland with the 109th Air Guard, I thank Renee Crain of the National Science Foundation. Meeka Kilabuk and Aaja Peter shared their knowledge of Nunavut at an Arctic Symposium at Bates College and the Chewonki Foundation. Dave Reid of Polar Sea Adventures of Pond Inlet, Baffin, provided logistics for many trips to the floe edge.

A few years ago, while my photography was on exhibit at the LL Bean store in Maine, Dr. Edward Morse introduced himself and revealed the source of his familiarity with my subject. He had been the medical doctor on the last Arctic Expedition conducted by Donald B. MacMillan, and both of his wives, sisters Helga and Inger Knudsen, had been raised in Ilulissat, or Jakobshavn as it was then called as a Danish colony, where their father was governor of Northwest Greenland. Ed and Inger have shared reminiscences of explorers, artists, and family, that have brought back to life the past of northwestern Greenland.

Ronald Levere, Digital Imaging Consultant, has my gratitude for sharing his voluminous knowledge of computers and cameras. Richard D. Kelly Jr. initiated the book's maps, which artist Marcia Bakry of the Smithsonian's Department of Anthropology enhanced and completed. Ms. Bakry originated the map of Maine and was gracious in making modifications as our knowledge of place-names grew. Laura Fleming of the Arctic Studies Center at the Smithsonian also provided logistical support in preparing artwork for publication.

Early in the process of digitally producing components for a book, I met Kate Carpenter, Training Manager at Burgess Computer in Bath, Maine, who helped immensely with the finer points of composing digital documents. Appreciation is also extended to Todd Kent of Xpress Copy of Portland, Maine, for running draft copies when the manuscript was under revision.

In Greenland, when Lindsay and I were invited to a birthday party in Uummannaq, we met by chance educators Kunuunnguaq Fleischer and his wife Naja Rosing. In subsequent travels to Greenland, Kunuunnguaq found settlements in which I, or we, would be posted. It was through one of those postings that I came to Uummannaq and met Ann Andreasen, Director of the Uummannaq Children's Home. Never before have I met anyone as insightful and as generous as Ann. Uummannaq, a town adjacent to Ilulissat, is now my home away from home, and Ann has become a frequent and welcome visitor to Maine. Through Ann I met her assistant, René Kristensen, and collaborated with them to create student exchanges between Maine and Uummannaq; subsequently Bowdoin College, the Chewonki Foundation, and other organizations have helped realize these educational interfaces. Through these exchanges, both personal and institutional, Maine's northern connections continue!

In Harrington Harbor, Bill and I were fortunate to meet the Reverend Bob Bryan, Archdeacon of the Anglican Church on Quebec's Lower North Shore and a co-founder of the Downeast humor team in Maine known as "Bert and I." Larry Morris, President of the Quebec-Labrador Foundation that Rev. Bryan founded, has continued this institution's invaluable support. Firm and generous support for this book has also come from the Maine Community Foundation, which provided grants from the Marshall Dodge Fund. To Burnham & Morrill Company of Portland and Janis Astle: thank you for the annual contribution of food-stuffs for the larder of the *Pitsiulak.*

A special note of thanks is extended to His Excellency Friis Arne Petersen, former Ambassador of the Royal Danish Embassy to the United States, for a financial grant and for his interest in our work. Our appreciation is extended to the Maine Arts Commission for a grant supported by the National Endowment for the Arts.

Numerous colleagues have assisted with this project: we are especially grateful to Susan Kaplan of the Peary-MacMillan Arctic Museum at Bowdoin College and to Walter Adey and Stephen Loring of the Smithsonian Institution for sharing their expertise to enhance the manuscript and illustrations. We thank Aqqaluk Lynge, Kununnguak Fleischer, and Danish PhD student Jes Harfield for reviewing drafts for accuracy and orthography, without prejudice for any errors for which authors bear full responsibility.

The editing and design skills of Tish O'Connor and Dana Levy of Perpetua Press, Santa Barbara, have transformed an array of material gathered from seemingly disparate sources and our many travels into a text that is coherent and a design that enhances its content. Thank you, Tish and Dana! The decision by Smithsonian Books to issue this publication is most appreciated.

Abram, David
1996 *The Spell of the Sensuous.* New York: Vintage Books.

ACIA (Arctic Climate Impact Assessment)
2004 *Impacts of a Warming Arctic: Arctic Climate Impact Assessment.* Cambridge: Cambridge University Press. http://www.acia.uaf.edu.

Alley, Richard B.
2002 *The Two-Mile Time Machine. Ice Cores, Abrupt Climate Change, and Our Future.* Princeton: Princeton University Press.
2004 Abrupt Climate Change. *Scientific American* (November: 62–69).
2011 *Earth – The Operator's Manual.* New York: W. W. Norton Company, Inc.

Alley, R.B., P.A. Mayewski, T. Sowers, M. Stuiver, K.C. Taylor and P.U. Clark
1997 Holocene Climatic Instability: A Prominent, Widespread Event 8,200 Years Ago. *Geology* 25(6): 483–486.

Alley, Richard B., and Anna Maria Ágústsdóttir
2005 The 8k Event: Cause and Consequences of a Major Holocene Abrupt Climate Change. *Quaternary Science Reviews* 24 (10–11): 1123–49.

Alley, Richard B., and Jan Joughin
2012 Perspective - Climate Change: Modeling Ice-Sheet Flow. *Science* 336 (6081): 551–552.

Allin, Lawrence C., and Richard W. Judd
1995 Creating Maine's Resource Economy, 1783–1861. In *Maine – The Pine Tree State from Prehistory to the Present*, edited by Richard W. Judd, Edwin A. Churchill, and Joel W. Eastman, 275–280. Orono, ME: University of Maine Press.

Alsford, Stephen (ed.)
1993 *The Meta Incognita Project. Contributions to Field Studies.* Mercury Series Directorate Paper 6. Ottawa and Washington, DC: Canadian Museum of Civilization and Arctic Studies Center, National Museum of Natural History, Smithsonian Institution.

Andersen, Mogens Voigt
1998 *Ilulissat Katersugaasiviat / Jakobshavn Museum.* Ilulissat, Greenland: GrønlandsAnnoncebureauApS.

Anderson, Alan
2009 *After the Ice: Life, Death, and Geopolitics in the New Arctic.* New York: HarperCollins for Smithsonian Books.

Aporta, Claudio
2009 The Trail as Home: Inuit and Their Pan-Arctic Network of Routes. *Human Ecology.* Published online by Springer Science and Business Media 37 (January 23): 131–146.

Aporta, Claudio, and Eric Higgs
2005 Satellite Culture: Global Positioning Systems, Inuit Wayfinding, and the Need for a New Account of Technology. *Current Anthropology* (December) 46 (5): 729–754.

Appenzeller, Tim
2007 The Big Thaw. *National Geographic* (June): 56–71.

Arctic Council
2004 *Impacts of a Warming Arctic Highlights.* Fairbanks, AK: International Arctic Research Center.

Arctic Studies Center newsletter
2005 Number 13 (Dec). Narwhals, Inuit and IPY. National Museum of Natural History, Smithsonian Institution.

Armitage, Peter
1992 Religious Ideology Among the Innu of Eastern Quebec and Labrador. *Religiologiques* 6: 64–110.
1991 *The Innu.* New York: Chelsea House Publishers.

Arneborg, Jette
2000 Greenland and Europe. In *Vikings: The North Atlantic Saga*, edited by William Fitzhugh and Elizabeth I. Ward, 304–317. Washington, DC: Smithsonian Institution Press.

Arneborg, J., J. Heinemeier, N. Lynnerup, H.L. Nielsen, N. Rud, and A.E. Sveinbjörnsdóttir
1999 Change of diet of the Greenland Vikings Determined from Stable Carbon Isotope and 14C Dating of their Bones. *Radiocarbon* 41(2): 157–168.

Atlantic Geoscience Society.
2001 *The Last Billion Years.* Special Publication No. 15. Halifax, NS: Nimbus Publishing.

Audubon, John James
1827–1838 *The Birds of America, from Drawings Made in the United States and Their Territories.* London: Published by the Author.

1897 *Audubon and His Journals,* by Maria R. Audubon. With Zoological and Other Notes by Elliott Coues. New York: Scribner's Sons.

Auger, Reginald, William Fitzhugh, Lynda Gullason, and Ann Henshaw
1995 Material Evidence from the Frobisher Voyages: Anglo-Inuit Contact in the North American Arctic in the late Sixteenth Century. In *Trade and Discovery: The Scientific Study of Artifacts from Post-Medieval Europe and Beyond*, edited by Duncan R. Hook and David Gaimster, 13–28. London: British Museum Occasional Paper 109.

Bamber, Jonathan, Michiel van den Broeke, Janneke Ettema, Jan Lenaerts, and Eric Rignot
2012 Recent Large Increases in Freshwater Fluxes from Greenland into the North Atlantic. *Geophysics Research Letters*, 39, L19501, doi:10.1029/2012GL052552.

Barcott, Bruce.
2011 Arctic Fever. *Onearth* (Spring): 28–37.

Barkham, Selma Huxley
1980 Notes on the Strait of Belle Isle During the Period of Basque Contact with Indians and Inuit. *Etudes/Inuit/Studies* 4 (1–2): 51–58.
1984 The Basque Whaling Establishments in Labrador 1536–1632. A Summary. *Arctic* 37 (4): 515–519.
1989 *The Basque Coast of Newfoundland.* Plum Point, NL: Great Northern Peninsula Development Corporation.

Barrios, Greg
2003 Barry Lopez the Acclaimed Writer Reflects on Nature and Culture. *Nature Conservancy* (Winter).

Barnett, Tim P., David W. Pierce, and Reiner Schnur
2001 Detection of Anthropogenic Climate Change in the World's Oceans. *Science* 292 (5515): 270–274. April 13.

Bartlett, Robert A., Captain

1928 *The Log of Bob Bartlett*. New York: G. P. Putnam's Sons.

Bates, Robert L., and Julia Jackson (eds.)

1984 *Dictionary of Geological Terms*. (3rd ed). New York: Anchor Books.

Bath, Richard C.

1975/1976 The Status of Borderlands Studies: Political Science. *The Social Science Journal* 12–13: 55–67.

Baxter Marlow, Connie

1999 [1972] *Greatest Mountain: Katahdin's Wilderness*. Gardiner, ME: Tilbury House.

Beattie, Owen, and John Geiger

1987 *Frozen in Time: The Facts of the Franklin Expedition*. Vancouver, BC: Douglas & McIntyre.

Beckenstein, Myron

2004 Maine's Lost Colony. *Smithsonian* magazine (February): 18–19.

Belvin, Cleophas

2006 *The Forgotten Labrador: Kegaska to Blanc-Sablon*. Montreal: McGill-Queens University Press.

Bergerud, A.T., Stuart N. Luttich, and Lodewijk Camps (eds.)

2008 *The Return of Caribou to Ungava*. Montreal: McGill-Queen's University Press.

Best, George

1578 *A true discourse of the late voyages of discoverie, for the finding of a passage to Cathaya, by the northwest, under the conduct of Martin Frobisher general*. London. Reprinted in Stefansson and McCaskill, 1938.

Bogoyavlenskiy, Dmitry, and Andy Siggner

2004 Arctic Demography. In *Arctic Human Development Report* (November). Fairbanks, AK: The Arctic Council.

Bolster, W. Jeffrey

2012 *The Mortal Sea: Fishing the Atlantic in the Age of Sail*. Cambridge, MA: The Belknap Press of Harvard University Press.

Boorstin, Daniel J.

1983 *The Discoverers: A History of Man's Search to Know His World and Himself*. New York: Vintage Books.

Borlase, Tim

1993 *The Labrador Inuit*. Labrador Studies. Happy Valley-Goose Bay: Labrador East Integrated School Board.

1994 *The Labrador Settlers, Métis, and Kablunâjuit*. Labrador: Labrador East Integrated School Board.

Bouchard, Serge

2006 *Caribou Hunter: A Song of Vanished Innu Life*. Vancouver: Grey Stone Books.

Bourque, Bruce J.

2012 *Swordfish Hunters. The History and Ecology of an Ancient American Sea People*. Boston: Bunker Hill Publishing.

Box, J.E., X. Fettweis, J.C. Strove, M. Tedesco, D.K. Hall, and K. Steffen

2012 *The Cryosphere* 6: 821–839.

Box, Jason E. http://www.meltdactor.org/blog/?p=762

2012 Continued Retreat of Greenland's Most Productive Glacier. Melt Factor Blog, Sept. 5, 2012

Branan, Nicole

2009 Ocean's Conveyor Belt not Quite Complete. *Earth* (August): 21.

Berton, Pierre

1988 *The Arctic Grail: The Quest for the North West Passage and the North Pole 1818–1909*. Toronto, ON: McClelland and Stewart.

Brinkhoff, Thomas

2009 *Greenland: Municipalities, Major Towns & Settlements – Statistics and Maps on City Population*. Nuuk: Government of Greenland.

Broad, William J.

2005 It's Sensitive Really. *The New York Times*, Science D1, D4 , Dec. 13.

Brodo, Irwin M., Sylvia Duran Sharnoff, and Stephen Sharnoff

2001 *Lichens of North America*. New Haven, CT: Yale University Press.

Brody, Hugh

2000 *The Other Side of Eden: Hunters, Farmers, and the Shaping of the World*. New York: North Point Press.

Brown, Paul

2007 Melting Ice Cap Triggering Earthquakes. (Interview with Robert Corell, chairman of the Arctic Climate Impact Assessment). *The Guardian*, Sept. 7.

Brown, Patricia Leigh

2012 For Local Fisheries, a Line of Hope. *New York Times*, Oct. 1.

Bryce-Bennett, Carol

1977 *Our Footprints Are Everywhere: Inuit Land Use and Occupancy in Labrador*. Nain, Labrador: Labrador Inuit Association.

Buggey, Susan

1999 *An Approach to Aboriginal Cultural Landscapes*. Ottawa: Historic Sites and Monuments Board of Canada, Parks Canada.

2004 An Approach to Aboriginal Cultural Landscapes in Canada. In *Northern Ethnographic Landscapes: Perspectives from Circumpolar Nations*, edited by Igor Krupnik, Rachel Mason, and Tonia W. Horton, 17–44. Washington, DC: Arctic Studies Center, National Museum of Natural History, Smithsonian Institution.

Burnham, Dorothy K.

1992 *To Please the Caribou: Painted Caribou-Skin Coats Worn by the Naskapi, Montagnais, and Cree Hunters of the Quebec-Labrador Peninsula*. Toronto: Royal Ontario Museum.

Cabot, William Brooks

1912 *In Northern Labrador*. London: J. Murray.

1920 *Labrador*. Boston: Small, Maynard and Company.

Campbell, David G.

1992 *The Crystal Desert: Summers in Antarctica*. New York: Houghton Mifflin Company.

Canadian Geographic

2012 Canadian Scientists Louis Fortier and Serge Payette Received the Weston Family Prize for Lifetime Achievement in Northern Research. December 12.

Carter, W. Hodding

2000 *A Viking Voyage: In Which an Unlikely Crew of Adventurers Attempts an Epic Journey to the New World*. New York: Ballantine Books.

Churchill, Edwin A.

1995 The European Discovery of Maine. In *Maine – The Pine Tree State from Prehistory to the Present*, edited by Richard W. Judd, Edwin A. Churchill, and Joel W. Eastman, 31–50. Orono, ME: University of Maine Press.

Clements, Robert Markham

1889 *A Life of John Davis, the Navigator, 1550–1605: Discoverer of Davis Straits*. New York: Dodd, Mead and Company.

Colgan, Charles

1982 The Maine Economy: A Forecast to 1990. Augusta, ME: State Planning Office.

Conkling, Philip, Richard Alley, Wallace Broecker, and George Denton

2011 *The Fate of Greenland—Lessons from Abrupt Climate Change*. Cambridge, MA: MIT Press.

Connor, Steve

2006 Scientists Condemn US as Emissions of Greenhouse Gases Hit Record Level. *The Independent*, April 20.

Cook, Frederick A.

2001 [1911] *My Attainment of the Pole*. New York: Cooper Square Press.

Cornwallis, Graeme, and Deanna Swaney

2001 *Iceland, Greenland & the Faroe Islands*. (4th ed.). Victoria, Australia: Lonely Planet Publications.

Corson, Trevor

2002 Centerpiece: Stalking the American Lobster. *The Atlantic* (April): 61–81.

Cox, Steven L.

2000 A Norse Penny from Maine. In *Vikings: The North Atlantic Saga*, edited by William W. Fitzhugh and Elisabeth I. Ward, 206–207. Washington, DC: Smithsonian Institution Press.

Cronon, William

1983 *Changes in the Land: Indians, Colonists, and the Ecology of New England*. New York: Hill and Wang.

Crumlin-Petersen, Ole

2010 *Archaeology and the Sea in Scandinavia and Britain*. Maritime Culture of the North, 3. Roskilde, Denmark: Viking Ship Museum.

Crutzen, Paul

2002 The Geology of Mankind. *Nature* 415(6867): 23.

Cunliffe, Barry

2008 *Europe Between the Oceans: 9000 BC–AD 1000*. New Haven, CT: Yale University Press.

Damas, David

1984 Introduction. In *Handbook of North American Indians*. Vol. 5, *Arctic*. David Damas, volume editor; William C. Sturtevant, general editor, vol. 1–7. Washington, DC: Smithsonian Institution.

D'Anglure, Saladin

1984 Inuit of Quebec. In *Handbook of North American Indians*. Vol. 5, *Arctic*. David Damas, volume editor; William S. Sturtevant, general editor, vol. 1–7, 476–507. Washington, DC: Smithsonian Institution.

Dansgaard, W., H.B. Clausen, N. Gundestrup, S.J. Johnsen, and C. Rygner

1985 Dating and Climatic Interpretation of Two Deep Greenland Ice Cores. In *Greenland Ice Cores: Geophysics, Geochemistry, and the Environment*, edited by C.C. Langway Jr., H. Oeschger, and W. Dansgaard. *Geophysics Monograph* 33: 71–76.

Dansgaard, W., J.W.C. White, and S.J. Johnsen

1989 The Abrupt Termination of the Younger Dryas Climate Event. *Nature* 339: 532–534.

Davidson, James W., and John Rugge

1988 *Great Heart: The History of a Labrador Adventure*. New York: Viking Press.

Davies, Douglas R., and Thomas D. Davies

1990 New Evidence Places Peary at the Pole. *National Geographic Magazine* 177 (1). January. www.pearyhenson.org/dougdavies/navigationreport2.htm.

Diamond, Jared

2005 *Collapse: How Societies Choose to Fail or Succeed*. New York: Viking.

2012 *The World Until Yesterday: What Can We Learn from Traditional Societies?* New York: Viking.

Dick, Lyle

2001 *Musk Ox Land: Ellesmere Island in the Age of Contact.* Calgary, AB: University of Calgary Press.

Dillon, John

2001 William Brooks Cabot: The Brief Life of a Gentleman-Explorer, 1858–1949. *Harvard Magazine,* September–October. http://harvardmagazine.com/2001/09/william-brooks-cabot.html

Dowdeswell, Julian, and Michael Hambrey

2002 *Islands of the Arctic.* Cambridge, UK: Cambridge University Press.

Duchacek, Ivo D.

1986 *The Territorial Dimension of Politics: Within, Among, and Across Nations.* Boulder, CO: Westview Press.

Duff, John Alton

2009 The Hague Line in the Gulf of Maine: Impetus or Impediment to Ecosystemic Regime Building. *Gulf of Maine Symposium* 15 (2): 285–292. Portland: University of Maine, School of Law.

Dumais, Pierre, and J. Poirier

1994 Témoinage d'un Site Archéologiques Inuit, Baie des Belles Amours, Basse-Côte-Nord (Evidence of an Inuit archaeological site at Baie des Belles Amours, Lower North Shore). *Recherches Amérindiennes au Québec* 23(1–2): 18–30.

Dyer, Gwynne

2000 *Full Circle: First Contact.* Kevin McAleese (ed.). St. John's, NL: Newfoundland Museum.

Dykstra, Monique

2001 *Alone in the Appalachians: A City Girl's Trek from Maine to the Gaspésie.* Vancouver, BC: Rain Coast Publishers.

Eberhart, M.J.

2000 *Ten Million Steps: The Epic Trek of the Nimblewill Nomad.* Helena, MT: SkyHouse Publishers.

Economist (The)

2012 *Arctic Politics Amid the Thaw,* March 24: 61.

Ehrlich, Gretel

2001 *This Cold Heaven: Seven Seasons in Greenland.* New York: Pantheon Books.

Eicken, Hajo

2010 Indigenous Knowledge and Sea Ice Science: What We Can Learn from Indigenous Ice Users? In *SIKU: Knowing Our Ice – Documenting Inuit Sea-Ice Knowledge and Use,* edited by Igor Krupnik, Claudio Aporta, Shari Gearheard, Gita J. Laidler, and Lene Kielsen Holm, 357–376. New York: Springer.

Eiseley, Loren

1957 [1946] *The Immense Journey.* New York: Vintage Books.

Fagan, Brian

2000 *The Little Ice Age.* New York: Basic Books.

2006 *Fish on Friday: Feasting, Fasting, and the Discovery of the New World.* New York: Basic Books.

Fallows, James

2008 China's Silver Lining. *The Atlantic* (June): 36–50.

Feldman, Stacy

2009 Bolivia's Chacaltaya Glacia Melts to Nothing Six Years Early. *Inside Climate News,* May 6.

Ferris, Scott R., and Ellen Pearce

1998 *Rockwell Kent's Forgotten Landscapes.* Camden, ME: Down East Books.

Fick, Steven, and Alyssa Julie

2008 Á la Carte: Slicing the Polar Pie. *Canadian Geographic* (Jan/Feb): 40–41.

Fischer, David Hackett

2008 *Champlain's Dream: The European Founding of North America.* New York: Simon & Schuster.

Fitzhugh, Lynne D.

1999 *The Labradorians: Voices from the Land of Cain.* St. John's, NL: Breakwater Press.

Fitzhugh, William W.

1972 *Environmental Archaeology and Cultural Systems in Hamilton Inlet, Labrador,* Smithsonian Contributions to Anthropology, 16. Washington, DC: Smithsonian Institution.

1975a Introduction. *Arctic Anthropology* 12(2): 1–6.

1975b A Maritime Arctic Sequence from Hamilton Inlet, Labrador. *Arctic Anthropology* 12(2): 117–138.

1984 Paleo-Eskimo Cultures of Greenland. In *Handbook of North American Indians. Vol. 5, Arctic.* David Damas, volume editor; William C. Sturtevant, general editor, vol. 1–7, 55–539. Washington, DC: Smithsonian Institution.

2000 Puffins, Ringed Pins, and Runestones: The Viking Passage to America. In *Vikings: The North Atlantic Saga,* edited by William W. Fitzhugh and Elisabeth I. Ward, 11–25. Washington, DC: Smithsonian Institution Press.

2006 Cultures, Borders, and Basques: Archaeological Surveys on Quebec's Lower North Shore. In *From the Arctic to Avalon: Papers in Honour of James A. Tuck,* edited by Lisa Rankin and Peter Ramsden, 53–70. British Archaeological Reports International Series 1507.

2008 Arctic and Circumpolar Regions. *Encyclopedia of Archaeology,* edited by Deborah M. Pearsall, 247–271. New York: Academic Press/Elsevier.

2009 Exploring Cultural Boundaries: the 'Invisible' Inuit of Southern Labrador and Quebec. In *On the Track of the Thule Culture from Bering Strait to East Greenland,* edited by Bjarne Grønnow, 129–148. Studies in Archaeology and History 15. Copenhagen: National Museum of Denmark.

2012 *The Gateways Project 2011: Land and Underwater Excavations at Hare Harbor, Mécatina.* Edited by William W. Fitzhugh. Underwater report by Erik Phaneuf. Photographic contributions by Wilfred Richard. Produced by Lauren Marr. Washington, DC: Arctic Studies Center, National Museum of Natural History, Smithsonian Institution.

2013 The Inuit of Southern Labrador and the Quebec Lower North Shore. In *Oxford Handbook of Arctic Archaeology,* edited by T. Max Friesen and Owen K. Mason. Oxford: Oxford University Press, forthcoming.

Fitzhugh, William W., and Henry F. Lamb

1985 Vegetation History and Culture Change in Labrador Prehistory. *Journal of Arctic and Alpine Research* 17(4): 357–370.

Fitzhugh, William W., and Jacqueline S. Olin

1993 *Archaeology of the Frobisher Voyages.* Washington: Smithsonian Institution Press.

Fitzhugh, William W., and Elisabeth I. Ward (eds.)

2000 *Vikings: The North Atlantic Saga,* Washington, DC: Smithsonian Institution Press.

Fitzhugh, William W., and Matthew D. Gallon

2002 *The Gateways Project 2002: Surveys and Excavations from Petit Mécatina to Belles Amours.* Washington, DC: Arctic Studies Center, National Museum of Natural History, Smithsonian Institution.

Fitzhugh, William W., Anja Herzog, Sophia Perdikaris, and Brenna McLeod

2011 Ship to Shore: Inuit, Early Europeans, and Maritime Landscapes in the Northern Gulf of St. Lawrence. In *The Archaeology of Maritime Landscapes: When the Land Meets the Sea,* edited by Ben Ford, 99–128. New York: Springer.

Fitzhugh, William W., and Eric Phaneuf

2012 Inuit Archaeology on the Quebec Lower North shore 2011. *Provincial Archaeology Office 2011 Archaeology Review* 10: 63–76.

Flannery, Tim

2005 *The Weather Makers.* New York: Atlantic Monthly Press.

Fleming, Fergus

2001 *Ninety Degrees North: The Quest for the North Pole.* New York: Grove Press.

Fobert, Emily

2007 Pole Positioning. *Canadian Geographic* (Jan/Feb): 24.

Folger, Mark, and Peter Essick

2010 Viking Weather. *National Geographic* (June) 217:6: 48–67.

Forbes, Bruce C.

2010 Who are the Peoples of the North? *The Circle* 2: 6–10.

Fortey, Richard

2004 *Earth: An Intimate History.* New York: Alfred A. Knopf.

Freuchen, Peter

1953 *Vagrant Viking: My Life and Adventures.* New York: Julian Messner, Inc.

2002 [1935] *Arctic Adventure: My Life in the Frozen North.* Guilford, CT: Globe Pequot Press.

Frenette, Pierre

1996 *Histoire de la Côte-Nord* (History of the North Shore). Collection Les Régions 9 du Québec. Institut Québecois de Recherches sur la Culture. Laval: Les Presses de l'Université Laval.

Frumhoff, Peter C., James J. McCarthy, Jerry M. Melillo, Susanne C. Moser, and Donald J. Wuebbles (eds.)

2007 *Confronting Climate Change in the U.S. Northeast – Science, Impacts, and Solutions.* Synthesis report of the Northeast Climate Impacts Assessment (NECIA). Cambridge, MA: Union of Concerned Scientists.

Funder, Svend

2001 Greenland During the Ice Ages. In *The Ecology of Greenland,* 48–54. Nuuk, Greenland: Ministry of Environment and Natural Resources.

Garreau, Joel

1981 *The Nine Nations of North America.* New York: Avon Books.

Génsbøl, Benny

2004 *Nature and Wildlife Guide to Greenland.* Copenhagen, DK: Gyldendal Publishers.

Gildersleeve, Charles R.

1975–1976 The Status of Borderlands Studies: Geography. *The Social Science Journal* 12/13: 19–28.

Goddard, Ives

1984 Synonymy by David Damas. In *Handbook of North American Indians.* Vol. 5, Arctic. David Damas, volume editor; William C. Sturtevant, general editor, vol. 1–7. Washington, DC: Smithsonian Institution.

Goldenberg, Suzanne

2012 Greenland Ice Sheet melted at unprecedented rate during July. *The Guardian,* July 24.

Goodnough, Abby

2011 Scientists Say Cod is Scant; Nets Say Otherwise. *New York Times,* Dec. 10.

Gore, Albert W.

2006 *An Inconvenient Truth: The Planetary Emergency of Global Warming and What We Can Do About It.* Emmaus, PA: Rodale Press.

Gorman, James
2012 How Brown and Polar Bear Split Up but Continued Coupling. *New York Times*, July 23.

Gosling, William Gilbert
1911 *Labrador: Its Discovery, Exploration, and Development*. New York: John Lane Company.

Goudie, Elizabeth
1973 *Woman of Labrador*. Edited and with an introduction by David Zimmerly. Toronto: Peter Martin Press.

Govier, Katherine
2003 *Creation: a Novel*. Woodstock, NY: Overlook Press.

Gray, J. S.
2002 Biomagnification in Marine Systems: The Perspective of an Ecologist. *Marine Pollution Bulletin* 45(1–12): 46–52.

Gregoire, George
2012 *Walk With My Shadow: The Life of an Innu Man*. St. John's: Creative Publishers.

Grenfell, Wilfred Tomason
1909 *Labrador: The Country and the People*. New York: Macmillan.

1990 [1909] Adrift on an Ice Pan. William Pope (ed.). In *The Best of Wilfred Grenfell*, 17–38. Hantsport, NS: Lancelot Press.

Grenier, Robert, Marc-André Bernier, and Willis Stevens (eds.)
2007 *The Underwater Archaeology of Red Bay: Basque Shipbuilding and Whaling in the 16th Century*. 5 volumes. Ottawa: Parks Canada.

Guest, A.
2011 The Fluted Point of Ramah Chert. *Mammoth Trumpet* 26(4): 15–18.

Gurney, Alan
2004 *Compass: A Story of Exploration and Innovation*. New York: W.W. Norton.

Gulløv, Hans-Christian
2001 Natives and Norse in Greenland. In *Vikings: the North Atlantic Saga*, edited by William Fitzhugh and Elisabeth Ward, 319–326. Washington, DC: Smithsonian Institution Press.

Haake, Katharine
2002 *That Water, Those Rocks*. Western Literature Series. Reno: University of Nevada Press.

Hadingham, Evan
2012 America's First Immigrants. Reprint of *Smithsonian* magazine (Nov. 2004) in *United States History Annual Editions*. Robert James Maddox (ed.). Unit 1: 1–3.

Hagler, G.S.W., M. H. Bergin, E. A. Smith, M. Town, J.E. Dibb
2008 Local Anthropogenic Impact on Particulate Elemental Carbon Concentrations at Summit Greenland. *Atmospheric Chemistry and Physics* 8: 2485–2491.

Hall, Charles Francis
1865 *Arctic Researches and Life Among the Eskimaux. Being the Narrative of an Expedition in Search of Sir John Franklin in the Years 1860, 1861, and 1862*. New York: Harper.

Hallendy, Norman
2008 *Inuksuit: Silent Messengers of the Arctic*. Vancouver, BC: Douglas & McIntyre.

Harp, Elmer Jr.
2003 *Lives and Landscapes: a Photographic Memoir of Outport Newfoundland and Labrador, 1949–1963*. Edited and introduced by M.A.P. Renouf, with a contribution by Elaine Groves Harp. Montreal: McGill-Queen's University Press.

Hart Hansen, Jens Peder, Jørgen Meldgaard, and Jørgen Nordqvist (eds.)
1991 [1985] *The Greenland Mummies*. London: British Museum Press.

Hastrup, Kirsten
2007 Thule: Anthropology and the Call of the Unknown. *Journal of the Royal Anthropological Society* (n.s.) 798–799: 801; [orig. Søby, R.M. 1983. Knud Rasmussens tale vedprocessen I Haag 1932: indledningved R.M. Søby. *Grønland*, 177–181.]

Hattersley-Smith, Geoffrey, and Parks Canada
1998 *Geographical Names of the Ellesmere Island National Park Reserve and Vicinity*. Calgary, AB: University of Calgary, Arctic Institute of North America.

Hawkes, Ernest W.
1916 *The Labrador Eskimo*. Ottawa: Department of Mines, Geological Survey, No.14, Memoir 91.

Heat-Moon, William Least
1991 *PrairyErth*. Boston: Houghton Mifflin.

Heinrich, Bernd
2003 *Winter World – The Ingenuity of Animal Survival*. New York: HarperCollins.

2007 *The Snoring Bird: My Family's Journey Through a Century of Biology*. New York: HarperCollins.

Heide-Jørgensen M. P., K. L. Laidre, D. Borchers, T. A. Marques, H. Stern, and M. J. Simon
2010 The Effect of Sea Ice Loss on Beluga Whales (*Delphinapterus leucas*) in West Greenland. *Polar Research* 29: 198–208.

Heide-Jørgensen, Mads Peter, Kristin L. Laidre, Lori T. Quakenbush, and John J. Citta
2011 The Northwest Passage Opens for Bowhead Whales. *Biology Letters* 8(2): 270–273 (doi: 10.1098/rsbl.2011.0731 Biol. Lett. rsbl20110731)

Henderson, Bruce
2005 *True North: Peary, Cook, and the Race to the Pole*. New York: W. W. Norton.

Henriksen, Georg
1973 *Hunters in the Barrens; The Naskapi on the Edge of the White Man's World*. Newfoundland Social and Economic Studies, 12. St. John's: Institute for Social and Economic Research, Memorial University of Newfoundland.

Herzog, Anja, and Jean-François Moreau
2004 Petit Mécatina 3: A Basque Whaling Station of the Early 17th Century? In *The Gateways Project 2004: Surveys and Excavations from Chevery to Jacques Cartier Bay*, edited by William W. Fitzhugh, Yves Chrétien, and Helena Sharp, 76–87. Washington, DC: Arctic Studies Center, National Museum of Natural History, Smithsonian Institution.

2006 European Glass Trade Beads, Neutron Activation Analysis, and the Historical Implications of Dating Seasonal Basque Whaling Stations in the New World. Zaragoza, Spain: Institución Fernando el Católico, Publication 2,621: 495–502.

Heyes, Scott A., and Kristofer M. Helgen
2013 *Mammals of Labrador and Ungava. The 1882–1884 Fieldnotes of Lucien M. Turner Together with Inuit and Innu Knowledge*. Washington, DC: Smithsonian Scholarly Press.

Hileman, Bette
2004 An Urgent Plea on Global Warming. *Chemical and Engineering News* (June 28): 44.

His Holiness The Dalai Lama.
1999 *Ethics for the New Millennium*. New York: Riverhead Books.

Holland, David M., Robert H. Thomas, Brad de Young, Mads H. Ribergaard, and Bjarne Lyberth
2008 Acceleration of Jakobshavn Isbrae Triggered by Warm Subsurface Ocean Waters. *Nature Geoscience* 1 (September 28): 659–664.

Holm, Bill
2007 *The Windows of Brimnes: An American in Iceland*. Minneapolis, MN: Milkweed Editions.

Holm, Lene Kielsen
2010 Sila-Inuk: Study of the Impacts of Climate Change in Greenland. In *SIKU: Knowing Our Ice – Documenting Inuit Sea-Ice Knowledge and Use*, edited by Igor Krupnik, Claudio Aporta, Shari Gearheard, Gita J. Laidler, and Lene Kielsen Holm, 145–160. New York: Springer.

Holt-Jensen, Arild
1988 *Geography: History and Concepts*. (2nd ed.). London, UK: Paul Chapman.

Horton, Tonia Woods
2004 Writing Ethnographic History: Historical Preservation, Cultural Landscapes, and Traditional Cultural Properties. In *Northern Ethnographic Landscapes: Perspectives from Circumpolar Nations*, edited by Igor Krupnik, Rachel Mason, and Tonia W. Horton, 65–80. Washington, DC: Arctic Studies Center, National Museum of Natural History, Smithsonian Institution.

House, John W.
1980 The Frontier Zone: A Conceptual Problem for Policy Makers. *International Political Science Review* 1: 456–477.

1981 Frontier Studies: An Applied Approach. In *Political Studies from Spatial Perspectives: Anglo-American Essays on Political Geography*, edited by Alan D. Burnett and Peter J. Taylor, 291–312. Toronto, ON: John Wiley and Sons.

1982 *Frontier on the Rio Grande. A Political Geography of Development and Social Deprivation*. Oxford, UK: Clarendon Press.

Houston, John
1995 *Confessions of an Igloo Dweller*. Boston: Houghton Mifflin Company.

Howat, Ian M., Ian Joughin, and Ted A. Scambos
2007 Rapid Change in Ice Discharge from Greenland Outlet Glaciers. *Science* (March 18) 315: 1559–1561.

Hubbard, Mina
1908 [2004] *A Woman's Way through Unknown Labrador*. London: Murray. Republished, Montreal: McGill-Queen's McGill University Press, 2004.

Huntford, Roland
2001 [1997] *Nansen: The Explorer as Hero*. London, UK: Abacus.

Huntington, Henry P., Shari Gearheard, and Lene Kielsen Holm
2010 The Power of Multiple Perspectives: Behind the Scenes of the Siku-Inuit-Hila Project. In *SIKU: Knowing Our Ice – Documenting Inuit Sea-Ice Knowledge and Use*, edited by Igor Krupnik, Claudio Aporta, Shari Gearheard, Gita J. Laidler, and Lene Kielsen Holm, 257–274. New York: Springer.

Ingstad, Anne Stine
1977 *The Discovery of a Norse Settlement in America. Excavations at L'Anse aux Meadows, Newfoundland 1961–1968*. Oslo: Universiteltforiaget.

Ingstad, Helge
1966 *Land Under the Polar Star*. New York: St. Martin's Press.

International Grenfell Association
2007 *Strategic Plan – Continuing the Mission*. International Grenfell Association. (www.iga.nf.net/IGA%20Strategic%20Plan%20Sept%202010.pdf).

Intergovernmental Panel on Climate Change (IPCC)

2007 *Climate Change 2007: The Physical Science Basis - Summary for Policy Makers*, February. http://www.ipcc.ch/publications_and_data/publications_and_data_reports.shtml

2011 Special Report on Managing the Risks of Extreme Events and Disasters to Advance Climate Change Adaptation (SREX). *IPCC SREX Summary for Policy Makers*. First Joint Session of Working Groups I and II. November 18. (ipcc-wg2.gov-SREX-images-uploads-SREX-SPM_Approved-HiRes_opt.pdf.webloc.).

Jenkins, Mark, and James Balog

2010 Melt Zone. *National Geographic* (June) 217: 6, 34–47.

Johnson, Brian Fisher

2010 Greenland: Energy Gold Mine or Empty Rock. *Earth* (April): 14.

Johnson, Samuel

2001–05 [1784] *Johnsonia*. Piozzi. Columbia Encyclopedia. 6th ed.: 154. New York: Columbia University Press.

Johnston D.W., Matthew T. Bowers, Ari S. Friedlaender, and David M. Lavigne

2012 The Effects of Climate Change on Harp Seals (*Pagophilus groenlandicus*). *PLoS ONE, 7(1): e29158 DOI: 10.1371/journal.pone.0029158.*

Johnston, D.W., Ari S. Friedlaender, L.G. Torres, and David M. Lavigne

2005 Variation in Sea Ice Cover on the East Coast of Canada from 1969 to 2002. Climate Variability and Implications for Harp and Hooded Seals. *Climate Research* 29: 209–222.

Joughlin, Ian, Waleed Abdalati, and Mark Fahnestock

2004 Large Fluctuations on Greenland's Jakobshavn Isbrae Glacier. *Nature* 432: 608–610.

Kaplan, Robert D.

2010 *Monsoon: The Indian Ocean and the Future of American Power.* New York: Random House.

2012 *The Revenge of Geography – What the Map Tells Us About Coming Conflicts and the Battle Against Fate.* New York: Random House.

Kaufman, Darrell S. David P. Schneider, Nicholas P. McKay, Caspar M. Ammann, Raymond S. Bradley, Keith R. Briffa, Gifford H. Miller, Bette L. Otto-Bliesner, Jonathan T. Overpeck, and Bo M. Vinther

2009 Recent Warming Reverses Long-Term Arctic Cooling. *Science* 325 (5945): 1236–1239.

Kavenna, Joanna

2005 *The Ice Museum: In Search of the Lost Land of Thule.* New York: Viking.

Kehrt, Christian.

2013 "The Wegener Diaries: Scientific Expeditions Into the Eternal Ice." Digital exhibition, Environment & Society Portal (Rachel Carson Center for Environment and Society, 2013). http://www.environmentandsociety.org/exhibitions/wegener-diaries

Kemmis, Daniel

1990 *Community and the Politics of Place.* Norman: University of Oklahoma Press.

Kent, Rockwell

1996 [1930] *N by E.* Hanover, NH: University Press of New England.

Knudsen, Pauline K., and Claus Andreasen

2009 *Cultural Historical Significance on Areas Tasersiaq and Tasersua in West Greenland & Suggestions for Salvage Archaeology and Documentation in Case of Damming Lakes.* Nuuk: Greenland National Museum and Archives.

Knudsen, Per, and Inger Holm Knudsen

2003/2006 Documentation dated March 4, 2003; notes of conversations of January 28, February 6, 15, and March 3, 2006; private Knudsen family archives, Owl's Head, Maine.

Kobalenko, Jerry

2002 *The Horizontal Everest.* New York: Soho.

2007 Between Nanuk and the Cold Grey Sea. *Canadian Geographic* (May/June): 38–52.

Kofoed, Emilie

2012 Nuna Minerals Maintains 2011 Expectation. *Greenland Oil & Minerals* 2. Nuuk: Sermetsiaq AG.

Kohlmeister, Benjamin, and George Kmoch

1814 *Journal of a Voyage from Okkak, on the Coast of Labrador, to Ungava Bay, Westward of Cape Chudleigh; Undertaken to Explore the Coast, and Visit the Esquimaux in That Unknown Region.* London: W. M'Dowall.

Kolbert, Elizabeth

2006 *Field Notes from a Catastrophe: Man, Nature and Climate Change.* New York: Bloomsburg.

Krupnik, Igor, Claudio Aporta, Shari Gearheard, Gita J. Laidler, and Lene Kielsen Holm (eds.)

2010 *SIKU: Knowing Our Ice – Documenting Inuit Sea-Ice Knowledge and Use.* New York: Springer.

Krupnik, Igor, and Dyanna Jolly

2010 Introduction: The Earth is Even Faster Now (Preface to the 2010 edition). In *The Earth is Faster Now – Indigenous Observations of Arctic Environmental Change*, edited by Igor Krupnik and Dyanna Jolly, xxiii–xxxv. Fairbanks, AK: Arctic Research Consortium of the United States.

Krupnik, Igor, Michael A. Lang, and Scott E. Miller (eds.)

2007 *Smithsonian at the Poles.* Washington, DC: Smithsonian Scholarly Press.

Krupnik, Igor, Rachel Mason, and Tonia W. Horton (eds.)

2004 *Northern Ethnographic Landscapes: Perspectives from Circumpolar Nations.* Washington, DC: Arctic Studies Center, National Museum of Natural History, Smithsonian Institution.

Krupnik, Igor, Rachel Mason, and Susan Buggey

2004 Introduction to Landscapes, Perspectives, and Nations. In *Northern Ethnographic Landscapes: Perspectives from Circumpolar Nations*, edited by Igor Krupnik, Rachel Mason, and Tonia W. Horton, 1–13. Washington, DC: Arctic Studies Center, National Museum of Natural History, Smithsonian Institution.

Kuhnlein, Harriet V., and Olivier Receveur

2007 Local Cultural Animal Food Contributes High Levels of Nutrients for Arctic Canadian Indigenous Adults and Children. *The Journal of Nutrition* (April) 137: 1110–1114.

Kunstler, James Howard

1993 *The Geography of Nowhere: The Rise and Decline of America's Man-Made Landscape.* New York: Simon & Schuster.

Kupp, Jan, and Simon Hart

1976 The Dutch in the Strait of Davis and Labrador during the 17th and 18th Centuries. *Man in the Northeast* 11(Spring): 3–20.

Kurlansky, Mark

1997 *Cod: A Biography of the Fish that Changed the World.* New York: Walker & Company.

1999 *The Basque History of the World.* New York: Penguin Books.

Lacoste, Karine N., and Garry B. Stenson

2000 Winter Distribution of Harp Seals (*Phoca groenlandica*) off Eastern Newfoundland and Southern Labrador. *Polar Biology* 23: 805–811.

Lavigne, David M., and Kit M. Kovacs

1988 *Harps and Hoods: Ice-Breeding Seals of the Northwest Atlantic.* Waterloo, Ontario: University of Waterloo Press.

Leacock, Eleanor

1954 The Montagnais "Hunting Territory" and the Fur Trade. *American Anthropological Association Memoir* 78.

1969 The Montagnais-Naskapi Band. Anthropological series 84. *National Museum of Canada Bulletin* 228: 1–17. Ottawa: National Museum of Canada.

Loewen, Brad, and Vincent Delmas

2011 Les Occupations Basques dans le Golfe de Saint-Laurent, 1530–1760. Périodisation, Répartition Géographique et Culture Matérielle (Basque occupations in the Gulf of Saint Lawrence, 1530–1760: dating, geographic distribution, and material culture). *Archéologiques* 24: 23–55.

Loomis, Chauncey C.

2000 *Weird and Tragic Shores: The Story of Charles Frances Hall, Explorer.* New York: Random House.

Lopez, Barry

1986 *Arctic Dreams: Imagination and Desire in a Northern Landscape.* New York: Bantam.

1998 *About This Life: Journeys on the Threshold of Memory.* New York: Vintage.

Loring, Stephen

1995 *Oh Darkly Bright: the Labrador Journeys of William Brooks Cabot, 1899–1910.* Exh. cat. Middlebury, VT: Johnson Memorial Gallery, Middlebury College.

1997 On the Trail to the Caribou House: Some Reflections on Innu Caribou Hunters in Ntessinan (Labrador). In *Caribou and Reindeer Hunters of the Northern Hemisphere*, edited by Lawrence Jackson and Paul Thacker, 185–220. London: Avebury Press.

2001 Introduction. In *Ethnology of the Ungava District, Hudson Bay Territory*, by Lucien M. Turner. Washington, DC: Smithsonian Institution Press.

2002 'And They Took Away the Stones from Ramah': Lithic Raw Material Sourcing and Eastern Arctic Archaeology. In *Honoring Our Elders: A History of Eastern Arctic Archaeology*, edited by William Fitzhugh, S. Loring, and D. Odess, 163–185. Contributions to Circumpolar Anthropology 2. Washington, DC: Arctic Studies Center, National Museum of Natural History, Smithsonian Institution.

2008 At Home in the Wilderness: the Mushuau Innu and Caribou. In *The Return of Caribou to Ungava*, edited by A.T. Bergerud, Stuart N. Luttich and Lodewijk Camps, 123–134. Montreal: McGill-Queen's University Press.

2009 From Tent to Trading Post and Back Again– Smithsonian Anthropology in Nunavut, Nunavik, Nitassinan, and Nunatsiavut: The Changing IPY Agenda, 1882–2007. In *Smithsonian at the Poles: Contributions to International Polar Year Science*, edited by Igor Krupnik, Michael A. Lang, and Scott E. Miller, 115–128. Washington, DC: Smithsonian Institution Scholarly Press.

Loring, Stephen, and Arthur Spiess

2007 Further documentation regarding the former existence of the Grizzly Bear (*Ursus arctos*) in northern Quebec-Labrador. *Arctic* 60(1): 7–16.

Lovelock, James

2006 *The Revenge of Gaia: Earth's Climate Crisis & the Fate of Humanity.* New York: Basic Books.

Low, Albert Peter

1896 *Report on Explorations in the Labrador Peninsula Along the East Main, Koksoak, Hamilton, Manicuagan, and Portions of Other Rivers in 1892–93–94–95.* Ottawa: S. E. Dawson.

Lynge, Aqqaluk

2008 "The Wind from the South." *The Veins of the Heart to the Pinnacle of the Mind.* Montreal and Hanover, NH: International Polar Institute.

Lynnerup, Niels
2000 Life and Death in Norse Greenland. In *Vikings: The North Atlantic Saga*, edited by William W. Fitzhugh and Elisabeth I. Ward, 285–294. Washington, DC: Smithsonian Institution Press.

Mailhot, José, and Andrée Michaud
1965 *North West River: Étude Ethnographique*. Travaux Divers 7. Laval: Centre d'Études Nordiques, Laval University.

Maine, Department of Marine Resources.
2012 *Historical Maine Fisheries Landings Data*. Augusta, ME. http://www.maine.gov/dmr/rm/groundfish/index.htm

Maine, Office of Policy and Management
2012 Maine Economics and Demographics Program, 1970–2010.

Malakoff, David
2007 Uncovering Basques in Canada. *American Archaeology* 11(2): 12–17.

Malkin, Carol
2005 Interview with Martin Nweeia: Nothing but the Tooth. *New Scientist*, Feb. 5.

Malaurie, Jean
2003 [1990] *Ultima Thule: Explorers and Natives in the Polar North*. New York: W. W. Norton.

Mann, Charles C.
2005 *1491: New Revelations of the Americas Before Columbus*. New York: Alfred A. Knopf.

Marcott, Shaun A., Jeremy D. Shakun, Peter U. Clark, and Akan C. Mix
2013 A Reconstruction of Regional and Global Temperature for the Past 11,300 Years. *Science* 339: 1198–1201.

Matthiessen, Peter
2003 *End of the Earth - Voyages to Antarctica*. Washington, DC: National Geographic.

Mayewski, Paul Andrew, and Frank White
2002 *The Ice Chronicles: The Quest to Understand Global Climate Change*. Hanover, NH: University Press of New England.

McGhee, Robert
1966 *Ancient People of the Arctic*. Vancouver: University of British Columbia Press.
2002 *The Arctic Voyages of Martin Frobisher: an Elizabethan Adventure*. London: The British Museum.
2008 Aboriginalism and the Problems of Indigenous Archaeology. *American Antiquity* 73 (4): 579–598.

McGovern, Thomas H.
2000 The Demise of Norse Greenland. In *Vikings: The North Atlantic Saga*, edited by William W. Fitzhugh and Elisabeth I. Ward, 327–339. Washington DC: Smithsonian Institution Press.

McGrath, Melanie
2007 *The Long Exile: A Tale of Inuit Betrayal and Survival in the High Arctic*. New York: Alfred A. Knopf.

McLeod, Brenna, M. W. Brown, M. J. Moore, W. Stevens, S. H. Barkham, M. Barkham, and B. N. White
2008 Bowhead Whales, and Not Right Whales, Were the Primary Target of 16th–17th Century Basque Whalers in the Western North Atlantic. *Arctic* 61(1): 61–75.

McManus, Gary E., and Clifford H. Wood
1991 *Atlas of Newfoundland and Labrador*. St. John's, NL: Memorial University of Newfoundland.

McNair, Wesley
2003 *Mapping the Heart: Reflections on Place and Poetry*. Pittsburgh, PA: Carnegie Mellon University Press.

2004 Ship, Dream, Pond, Talk. In *A Place on Water: Essays*. Gardiner, ME: Tilbury House.

Meadows, Donella H., Dennis L. Meadows, Jørgen Randers, and William W. Behrens III
1972 *The Limits to Growth*. New York: Universe Books.

Meldgaard, Jørgen
1960 On the Formative Period of the Dorset Culture. In *Prehistoric Cultural Relations Between the Arctic and the Temperate Zones of North America*, edited by John M. Campbell, 92–95. Technical Paper 11. Montreal: Arctic Institute of North America.

Merton, Thomas
1961 *New Seeds of Contemplation*. New York: New Directions.

Miller, Gifford H., John R. Southon, Chance Anderson, Helgi Björnsson, Thorvaldur Thordarson, Aslaug Geirsdottir, Yafang Zhong, Darren J Larsen, Bette L Otto-Bliesner, Marika M Holland, David Anthony Bailey, Kurt A. Refsnider, and Scott J. Lehman
2012 Abrupt Onset of the Little Ice Age Triggered by Volcanism and Sustained by Sea-ice/Ocean Feedbacks. *Geophysical Research Letters* 39 (2).

Miller, Jr., G. Tyler
2000 *Living in the Environment*. New York: Brooks-Cole.

Montañez, Isabelle, G. S. (Lynn) Soreghan, and Walter S. Snyder
2006 Earth's Fickle: Lessons Learned from Deep-Time Ice Ages. *Geotimes* (March).

Moon, T., B. Joughin, B. Smith, and I. Howat
2012 21st-century Evolution of Greenland Outlet Glacier Velocities. *Science* 336: 576–578.

Morison, Samuel Eliot
1971 *The European Discovery of America: The Northern Voyages A.D. 500–1600*. New York: Oxford University Press.

Moritz, Richard E., Cecilia M. Bitz, and Eric J. Steig
2002 Rev. of polar science issue - Dynamics of Recent Climate Change in the Arctic. *Science* (August 30) 297: 1497–1502.

Morrissey, Donna
2005 *Sylvanus Now*. New York: W.W. Norton.

Mortensen, Inger Holbech (ed.)
2001 *The Ecology of Greenland*. Nuuk, Greenland: Ministry of Environment and Natural Resources.

Morton, Mary Caperton
2012 Arctic Humidity on the Rise. *Earth* (October): 14.

Mudie, Peta J., Andre Rochon, and Elisabeth Levac
2005 Decadal-Scale Sea Ice Changes in the Canadian Arctic and their Impacts on Humans During the Past 4,000 Years. *Environmental Archaeology* 10: 113–126.

Myall, James
2012 *Franco-Americans in Maine: Statistics from the American Community Survey*. Report prepared for the Franco-American Taskforce. September 26.

NASA
2008 *Fastest Glacier in Greenland Doubles Speed*. December 1. www.physorg.com/news2285.html.

National Museum of Greenland
2006 *Women's Traditional Festive Dress Clothing*. Nuuk, Greenland.

National Snow and Sea Ice Data Center
2012 Arctic Sea Ice extent settles at record seasonal minimum. September 19. http://nsidc.org/arcticseaicenews/

Nielsen, Svend Erik
2008 The Icefjord – A Cold Harsh Workplace. *Greenland Today* (April): 64–67.

Nunavut Handbook.
1998 Iqaluit. Nunavut: Nortext Multimedia.

Nweeia, Martin T., Frederick C. Eichmiller, Cornelius Nutarak, Naomi Eidelman, Anthony A. Giuseppetti, Janet Quinn, James G. Mead, Kaviqanguak K'issuk, Peter V. Hauschka, Ethan M. Tyler, Charles Potter, Jack R. Orr, Rasmus Avike, Pavia Nielsen, and David Angnatsiak
2007 Considerations of Anatomy, Morphology, Evolution, and Function for Narwhal Detention. In *Smithsonian at the Poles*, edited by. I. Krupnik, M.A. Lang, and S. E. Miller, 223–240. Washington, DC: Smithsonian Scholarly Press.

Oates, David
2003 *Paradise Wild: Reimagining American Nature*. Corvallis: Oregon State University Press.

Odess, Daniel, Stephen Loring, and William W. Fitzhugh
2000 Skraeling: First Peoples of Helluland, Markland, and Vinland. In *Vikings: The North Atlantic Saga*, edited by William W. Fitzhugh and Elisabeth I. Ward, 193–205. Washington, DC: Smithsonian Institution Press.

Oeschger, Hans, Willi Dansgaard, and Chester Langway
1985 *Greenland Ice Core: Geophysics, Geochemistry, and the Environment*. Geophysical Monograph Series 33. Washington, DC: American Geophyical Union.

Packard, Alphonse S.
1885 Notes on the Labrador Eskimo and Their Former Range Southward. *The American Naturalist* 19 (5): 471–481.

Painter, James
2007 Greenland Sees Bright Side of Warming. BBC News. Interview with Robert Corell, chairman of the Arctic Climate Impact Assessment. September 14.

Parks Canada
1994 *Ellesmere Island National Park Reserve*. Ottawa: Parks Canada.
2003 *Visiting Sirmilik National Park of Canada*. Ottawa: Parks Canada.

Pauly, Daniel, and Jay Maclean
2003 *In a Perfect Ocean: The State of Fisheries and Ecosystems in the North Atlantic Ocean*. Washington, DC: Island Press.

Pelto, Mauri
2008 Moulins, Calving Fronts and Greenland Outlet Glacier Acceleration. *Real Climate: Climate Science from Climate Scientists*. April 18.

Petersen, Robert
1997 On Smell of Forests in the Greenlandic Myths and Legends. In *Fifty Years of Arctic Research – Anthropological Studies from Greenland to Siberia*, edited by Rolf Gilberg and Hans Christian Gulløv, 243–248. Ethnographical Series 18. Copenhagen: Department of Ethnography, National Museum of Denmark.

Philander, S. George
1998 *Is the Temperature Rising? The Uncertain Science of Global Warming*. Princeton, NJ: Princeton University Press.

Pielou, E.C.
1991 *After the Ice Age: The Return of Life to Glaciated North America*. Chicago, IL: University of Chicago Press.

Pintal, Jean-Yves
1998 *Aux Frontières de la Mer: la Préhistoire de Blanc-Sablon* (At the edge of the sea: the prehistory of Blanc Sablon). Québec: Ministère de la Culture et Communications.

Pittaway, Kim

2007 The Peter Principle [Aaju Peter]. *More* (Sept): 68–72.

Postrel, Virginia

2006 In Praise of Chain Stores. *Atlantic Monthly* (Dec): 164–167.

Proulx, Jean-Pierre

1993 *Les Basques et les Pêches de la Baleine au Labrador au XVIe Siècle* (Basques and whale hunting in Labrador in the 16th century). Études en Archéologie, Architecture, et Histoire. Ottawa: Lieux Historiques Nationaux, Service de Parcs, Environment Canada.

2007 Basque Whaling in Labrador: an Historical Overview. In *The Underwater Archaeology of Red Bay: Basque Shipbuilding and Whaling in the 16th Century*, edited by Robert Grenier, et al., I: 25–96. Ottawa: Parks Canada.

Purvis, William

2000 *Lichens*. Washington, DC: Smithsonian Institution Press.

Puyjalon, Henri de

2007 *Récits du Labrador* (Tales of Labrador). Introduction, Notes, Chronology by Daniel Chartier. Montréal: Imaginaire / Nord.

Reich, Robert R.

2007 *Supercapitalism: The Transformation of Business, Democracy, and Everyday Life*. New York: Alfred A. Knopf.

Revkin, Andrew C.

2012 DNA Study Finds Deeper Antiquity of Polar Bear Species. *New York Times,* April 19.

Richard, Wilfred E.

1991 International Factors and International Planning: A Borderlands Case Study of Canada and the United States. Ph.D. diss., University of Waterloo, Ontario, Canada.

1997 The International Appalachian Trail: Spanning a Two-Nation Bioregion. *The International Journal of Wilderness* 3 (1): 33–38.

2007 Arctic and Wilderness Travel–Hosts and Guests: The Territory of Nunavut, Canada. In *Proceedings of the Eighth World Wilderness Congress. RMRS-P-49, Science and Stewardship to Protect and Sustain Wilderness Values*, edited by Alan Watson, Janet Sproull, and Liese Dean, 152–161. Fort Collins, CO: US Department of Agriculture, Forest Service.

Richert, Evan

2006 Land Use in Maine: From Production to Consumption. In *Changing Maine 1960–2010: Teaching Guide*, edited by Richard Barringer, 33–35. Portland, ME: University of Southern Maine, New England Environmental Finance Center.

Roberts, David

2003 *Four Against the Arctic*. New York: Simon & Schuster.

Robertson, Samuel

1843 Notes on the Coast of Labrador. *Transactions of the Literary and Historical Society of Québec* 4(1): 27–53.

Rogers, Edward S.

1964 The Eskimo and Indian in the Quebec-Labrador Peninsula. In *Le Nouveau-Québec: Contribution à l'Etude de l'Occupation Humaine*, edited by Jean Malaurie and Jacques Rousseau, 211–249. Paris: Mouton.

Rogers, Edward S., and Eleanor Leacock

1981 Montagnais-Naskapi. In *Handbook of North American Indians*. Vol. 6, *Subarctic*. June Helm, volume editor; William S. Sturtevant, general editor, vol. 1–7, 169–189. Washington, DC: Smithsonian Institution.

Rollmann, Hans

2002 *Labrador Through Moravian Eyes: 250 Years of Art, Photographs & Records*. St. John's, NF: Department of Tourism, Culture, and Recreation.

Rompkey, Ronald

2001 *Jesse Luther at the Grenfell Mission*. Montreal, PQ: McGill-Queen's University Press.

2009 *Grenfell of Labrador – A Biography*. Montreal, PQ: McGill-Queen's University Press.

Rowley, Graham, and Susan Rowley

1997 Igloolik Island Before and After Jørgen Meldgaard. In *Fifty Years of Arctic Research – Anthropological Studies from Greenland to Siberia*, edited by Rolf Gilberg and Hans Christian Gulløv, 269–276. Ethnographical Series 18. Copenhagen: Department of Ethnography, National Museum of Denmark.

Roy-Sole, Monique

2003 River Jam. *Canadian Geographic* (March/April). www.canadiangeographic.ca/magazine/ma03/mosaic.asp

Ruddiman, William F.

2005 *Plows, Plagues & Petroleum: How Humans Took Control of Climate*. Princeton, NJ: Princeton University Press.

Russell, Nathan

2007 The Agricultural Impact of Global Climate Change. *Geotimes*. (April): 30–34.

Ryden, Kent C.

2001 Natural Landscapes, Cultural Regions; or, What is Natural About New England? In *Landscape with Figures: Nature & Culture in New England*, 199–281. Iowa City: University of Iowa Press.

Sachs, Jeffrey D.

2008 *Common Wealth: Economics for a Crowded Planet*. New York: Penguin Press.

2011 *The Price of Civilization – Reawakening American Virtue and Prosperity*. New York: Random House.

Safina, Carl

2002 *Eye of the Albatross*. New York: Henry Holt.

2011 *The View from Lazy Point: A Natural Year in an Unnatural World*. New York: Henry Holt.

Sale, Richard

2002 *Polar Reaches: The History of Arctic and Antarctic Exploration*. Seattle, WA: Mountaineers Books.

2006 *A Complete Guide to Arctic Wildlife*. Buffalo, NY: Firefly Books.

Sanguya, Joelie

2010 Preface. In *SIKU: Knowing Our Ice – Documenting Inuit Sea-Ice, Knowledge and Use*, edited by Igor Krupnik, Claudio Aporta, Shari Gearheard, Gita J. Laidler, and Lene Kielsen Holm, ix–x. New York: Springer.

Sargeant, David E.

1991 *Harp Seals, Man, and Ice*. Canada Special Publication of Fisheries and Aquatic Sciences, 114. Ottawa: Department of Fisheries and Oceans.

Scherman, Katharine

1956 *Spring on an Arctic Island*. Boston, MA: Little, Brown and Company.

Schlederman, Peter

1996 *Voices in Stone*. Calgary, AL: Arctic Institute of North America, University of Calgary.

Seelye, Katharine Q., and Jess Bidgood

2013 Officials Back Deep Cuts in Atlantic Cod Harvest to Save Fishery. *New York Times*, Jan. 30.

Shreeve, James

2006 The Greatest Journey. *National Geographic* (March).

Sigurdsson, Gisli

2000 An Introduction to the Vinland Sagas. In *Vikings: The North Atlantic Saga*, edited by William W. Fitzhugh and Elisabeth I. Ward, 232–237. Washington, DC: Smithsonian Institution Press.

Simon, Alvah

1999 *North to the Arctic: A Year in the Arctic Ice*. Camden, NJ: McGraw-Hill.

Smith, Jesse, and Julia Uppenbrink

2001 Earth's Variable Climatic Past, Introduction to Special Issue. *Science* (April 27) 292: 657.

Solberg, Ole Martin

1907 Beitragezurvorgeschichtederost-Eskimo (A study of the prehistory of the eastern Eskimo). Christiania (Copenhagen): J. Dybwab.

Speiss, Arthur, and Stephen L. Cox.

1976 Discovery of the Skull of a Grizzly Bear in Labrador. *Arctic* 29(4): 194-200.

Speck, Frank G.

1935 *Naskapi: Savage Hunters of the Labrador Peninsula*. Norman: University of Oklahoma Press.

Stafford, Edward P.

1998 *Peary and His Promised Land*. Bailey Island, ME: Friends of Peary's Eagle Island.

Stanford, Dennis J., and Bruce Bradley

2012 *Across Atlantic Ice: the Origin of America's Clovis Culture*. Berkeley: University of California Press.

Statistics Canada

2006 *Nunavut Population Counts*. Ottawa.

2010 *Province and Territory Population Counts-* Estimates July 1.

Statistics Greenland

2003 *Greenland in Figures 2003*. Nuuk, Greenland: Greenland Home Rule Government.

2010 *Greenland in Figures 2010*. 7th revised edition. Nuuk: Government of Greenland.

Stearns, Winfrid Alden

1884 *Labrador: a Sketch of its Peoples, its Industries and its Natural History*. Boston: Lee and Shepard.

Stefansson, Vilhjalmur, and Elizabeth McCaskill (eds.)

1938 *The Three Voyages of Martin Frobisher in Search of a Passage to Cathay and India by the North-West, A.D. 1576– 8. From the Original Text of George Best*. 2 Vols. London: Argonaut Press.

Stenson, G. B., and B. Sjare

1997 Seasonal Distribution of Harp Seals, *Phoca groenlandica*, in the Northwest Atlantic.International Council Explor. Sea. Commercial Fleet ICESC.M. 1997CC:10 (Biology and Behavior II). 23 pp.

Stix, Gary

2008 Traces of a Distant Past. *Scientific American* (July): 56–63.

Stoddard, Ellwyn R.

1975/1976 The Status of Borderlands Studies: Sociology and Anthropology. *The Social Science Journal* 12/13: 29–54.

Story, G.M., W.J. Kirwin, and J.D.A.Widdowson (eds.)

1990 *Dictionary of Newfoundland English*. 2nd edition. Toronto: University of Toronto Press.

Strong, William Duncan

1930 A Stone Culture from Northern Labrador and its Relation to the Eskimo-Like Cultures of the Northeast. *American Anthropologist* 30: 126–144.

Swanson, Roger Frank

1978 *Intergovernmental Perspectives on the Canadian – U.S. Relationship*. New York: NYU Press.

Symons, T.H.B. (ed.)

1999 *Meta Incognita: A Discourse of Discovery*. Hull: Canadian Museum of Civilization.

Taagholt, Jørgen, and Jens Claus Hansen
2001 *Greenland–Security Perspectives.* Trans. Daniel Lufkin. Fairbanks, AK: Arctic Research Consortium of the United States.

Tanner, Vainö
1947 *Outlines of the Geography, Life, and Customs of Newfoundland-Labrador.* 2 vols. New York: Macmillan.

Taylor, Andrew
1955 *Geographical Discovery and Exploration in the Queen Elizabeth Islands, Canada.* Department of Mines and Technical Surveys, Geographical Branch. Ottawa: Queen's Printer.

Taylor, J. Garth
1974 *Labrador Eskimo Settlements of the Early Contact Period.* Ottawa, Canada: National Museum of Man, Publications in Ethnology 9.
1984 Historical Ethnography of the Labrador Coast. In *Handbook of North American Indians.* Vol. 5, *Arctic.* David Damas, volume editor, William S. Sturtevant, general editor, vol 1–7, 508–521. Washington, DC: Smithsonian Institution.

Tersis, Nicole, and Pierre Taverniers
2010 Two Greenlandic Sea Ice Lists and Some Considerations Regarding Inuit Sea Ice Terms. In *SIKU: Knowing Our Ice – Documenting Inuit Sea-Ice Knowledge and Use,* edited by Igor Krupnik, Claudio Aporta, Shari Gearheard, Gita J. Laidler, and Lene Kielsen Holm, 413–426. New York: Springer.

Thoreau, Henry David
1988 [1864] *The Maine Woods.* New York: Penguin Books.

Townsend, Charles W.
1917 In Audubon's Labrador. *The Auk* 34: 133–146.

Trudel, François
1980 Les Relations entre les Français et les Inuit au Labrador Méridional, 1660–1760 (French-Inuit relations in southern Labrador). *Etudes Inuit Studies* 4(1–2): 135–145.

Tuck, James A.
1971 An Archaic Cemetery at Port au Choix, Newfoundland. *American Antiquity* 36–3: 343–358.
1975 The Northeastern Maritime Continuum: 8,000 Years of Cultural Development in the Far Northeast. *Arctic Anthropology* 12(2): 139–147.

Tuck, James A., and Robert Grenier
1981 A 16th-Century Basque Whaling Station in Labrador. *Scientific American* 245(5): 180–188.
1989 *Red Bay, Labrador: World Whaling Capital, A.D. 1550–1600.* St. John's, NL: Atlantic Archaeology.

Turgeon, Laurier
1994 Vers une Chronologie des Occupations Basques du Saint-Laurent du XVIe –XVIIe Siècle: un Retour à l'Histoire (Toward a chronology of Basque occupations of the Saint Lawrence in the 16th and 17th century: an historical reckoning). *Recherches Amérindiennes au Québec* 24(3): 3–15.

Turnbull, C.J.
1976 The Augustine Site: *A Mound from the Maritimes. Archaeology of Eastern North America* 4: 50–62.

Turner Lucien M.
2001 [1894] *Ethnology of the Ungava District, Hudson Bay Territory.* With an Introduction by Stephen Loring. Washington, DC: Smithsonian Institution Press.

Urquhart, Thomas
2004 *For the Beauty of the Earth: Birding, Opera, and Other Journeys.* Washington, DC: Shoemaker & Hoard.

US Central Intelligence Agency
2010 *The World Fact Book: North America – Greenland.* Washington, DC.

US Dept of Commerce and National Oceanic and Atmospheric Administration.
1939/1940. *Coastline of the United States*

Vail, David, Mark Lapping, Wilfred Richard, and Michael Cote
1998 *Tourism and Maine's Future: Toward Environmental, Economic and Community Sustainability.* Augusta, ME: Maine Center for Economic Policy.

Van derVeen, C.J., J. C. Plummer, and L.A. Stearns
2011 Controls on the recent speed-up of Jakobshavn Isbrae, West Greenland. *Journal of Glaciology* 57(204) 770–782.

Vaughan, Richard
2001 [1994] *The Arctic–A History.* Gloucester, UK: Sutton Publishing.

Vickery, James B., and Richard W. Judd
1995 Traditional Industries in the Age of Monopoly, 1865–1930. In *Maine – The Pine Tree State from Prehistory to the Present,* edited by Richard W. Judd, Edwin A. Churchill, and Joel W. Eastman, 404–410. Orono, ME: University of Maine Press.

Vongraven, D., and E. Richardson
2011 Biodiversity-Status and Trends of Polar Bears. National Oceanographic and Atmospheric Administration Arctic Report Card 2011. *http://www.arctic.noaa.gov/reportcard*

Wadden, Marie
1991 *Nitassinan: The Innu Struggle to Reclaim Their Homeland.* Vancouver: Douglas and McIntyre.

Waldron, Florence Mae
2005 'I've Never Dreamed It Was Necessary to Marry!': Women and Work in New England French Canadian Communities, 1870–1930. *Journal of American Ethnic History* (Winter): 34–64.
2005 The Battle over Female (In)Dependence: Women in New England Quebecois Migrant Communities, 1870–1930. *Frontiers: A Journal of Women Studies* 26: 158.

Wallace, Birgitta Linderoth
1991 *L'Anse aux Meadows: Gateway to Vinland.* Acta Archeologica 61: 166–198
2000a The Viking Settlement at L'Anse aux Meadows. In *Vikings: The North Atlantic Saga,* edited by William W. Fitzhugh and Elisabeth I. Ward, 208–224. Washington, DC: Smithsonian Institution Press.
2000b An Archaeologist's Interpretation of the Vinland Sagas. In *Vikings: The North Atlantic Saga,* edited by William W. Fitzhugh and Elisabeth I. Ward, 225–231. Washington, DC: Smithsonian Institution Press.
2006 *Westward Vikings: The Saga of L'Anse aux Meadows.* St. John's, NL: Historic Sites Association of Newfoundland and Labrador.

Wallace, Birgitta, and William W. Fitzhugh
2000 Stumbles and Pitfalls in the Search for Viking America. In *Vikings: The North Atlantic Saga,* edited by William W. Fitzhugh and Elisabeth I. Ward, 374–384. Washington, DC: Smithsonian Institution Press.

Wallace, Dillon
1905 *The Lure of the Labrador Wild.* New York: Flemming H. Revell.
1907 *The Long Labrador Trail.* New York: Outing Publishing Company.

Walsøe, Per
2003 Goodbye Thule – *The Compulsory History of Relocation in 1953.* Viborg, DK: Nørhaven Book.

Wasserrab, Jutta
2008 World Races to Win Arctic Treasure Hunt. *Deutsche Welle.* Science Section. (January 13).

Waterman, Jonathan
2001 *Arctic Crossing.* New York: Alfred A. Knopf.

Weaver, Ray
2012 Greenland Split on Uranium Mining. *The Copenhagen Post.* July 4, 2012

Webber, Michael
2009 Breaking the Energy Barrier. *Earth* (September): 46–51.

Wenzel, George
1991 *Animal Rights, Human Rights: Ecology, Economy and Ideology in the Canadian Arctic.* Toronto: University of Toronto Press.

Wessels, Tom
2001 *The Granite Landscape.* Woodstock, VT: The Countryman Press.

White, Winston C.
2003 *Labrador: Getting Along in the Big Land!* St. John's, NL: Flanker Press.

Williams, Jr., Richard S., and Jane G. Ferrigno
1995 *Satellite Image Atlas of Glaciers of the World – Greenland. U.S. Geological Survey Professional Paper 1386-C.* Reston, VA: U.S. Geological Survey.

Wilson, Edward O.
1999 *Consilience: The Unity of Knowledge.* New York: Vintage.
2012 *The Social Conquest of Earth.* New York: Liveright Publishing Corporation (W.W. Norton & Company).
2008 Foreword. In *Common Wealth: Economics for a Crowded Planet,* by Jeffrey D. Sachs, xi–xiii. New York: Penguin Press.

Wilson, Roger (ed.)
1976 *The Land that Never Melts. Auyuittuq National Park.* Toronto: Peter Martin.

Winchester, Ed, and Wilfred E. Richard
2006 Northeast Passage. *AMC Outdoors* (April): 28–33.

Wright, Ronald
2005 *A Short History of Progress.* New York: Carroll & Graf.

Yalowitz, Kenneth S., James F. Collins, and Ross A. Virginia
2008 *The Arctic Climate Change and Security Policy Conference – Final Report and Findings.* Hanover, NH: Dartmouth College.

Index

Number in italics refers to picture or map.

266